# THE PHILOSOPHY OF SCIENCE
## OF
## A.S. EDDINGTON

# THE PHILOSOPHY OF SCIENCE

## OF

## *A. S. Eddington*

BY

JOHN W. YOLTON

PRÉFACE PAR
**F. GONSETH**
*Zürich*

MARTINUS NIJHOFF / THE HAGUE / 1960

ISBN 978-94-015-0401-0       ISBN 978-94-015-1007-3 (eBook)
DOI 10.1007/978-94-015-1007-3

# PRÉFACE

En 1956, l'Académie Internationale de Philosophie des Sciences couronnait, à Bruxelles, l'étude de M. Yolton, intitulée *The Philosophy of Science of Arthur S. Eddington*, étude dont le présent ouvrage est la reprise. Pourquoi la personne, l'oeuvre et les idées de l'illustre physicien anglais avaient-elles été désignées à l'attention des candidats à ce Prix? Quels enseignements l'Académie Internationale de Philosophie des Sciences en attendait-elle? N'espérait-elle pas que le rapport de la recherche philosophique à la recherche scientifique pourrait en être éclairé?

M. Yolton offre à l'un des membres de son jury l'occasion de s'en expliquer à cet endroit. Je m'en vais donc le faire pour ce qui me concerne personnellement, conformément aux vues que je défends moi-même depuis assez longtemps.

La vision du monde et de l'homme, du philosophe et celle du savant s'accordent-elles ou s'opposent-elles? Pour ma part, je ne mets pas en doute qu'il existe entre la science et la philosophie une relation qu'il importe de mettre en évidence, mais j'insiste en même temps sur le fait que cette relation est ambivalente. A la fois, elles les unit, elle les sépare. Elle les fait tour à tour se rapprocher et s'éloigner l'une de l'autre. La science et la philosophie sont nécessaires l'une à l'autre, mais cette nécessité est parfois celle de leur alliance et parfois celle de leur antagonisme. N'est-ce pas là, d'ailleurs, la forme obligée de tout véritable dialogue?

Dans ce dialogue, le philosophe est fondamentalement du côté du pouvoir discursif. C'est le pouvoir de se servir de la langue comme d'un moyen d'expression, le moyen d'une expression sur laquelle la pensée qui cherche l'objet et l'objet qui s'offre à la pensée puissent se rencontrer. C'est le pouvoir de produire des formes verbales et d'ordonner leur mise en jeu de telle façon que la pensée puisse les épouser pour y reconnaître, temporairement ou durablement, la forme de sa propre activité, la structure de l'activité dans laquelle elle est engagée, en même temps que la structure de l'expérience dans laquelle elle s'engage.

Le pouvoir discursif n'a pas attendu que le philosophe existât pour déployer ces trames verbales: il n'y a pas de langage humain dans lequel il ne soit à l'oeuvre. Les communautés humaines se succèdent, l'héritage des langues qu'elles se transmettent s'altère, se transforme, évolue: le pouvoir discursif subsiste, il est inaliénable.

Le pouvoir discursif n'est pas au service du seul philosophe. Il anime l'image poétique comme il porte l'effort descriptif. Le philosophe l'accepte comme le moyen même de sa recherche. Mais cela ne suffit pas encore pour caractériser le philosophe. La géométrie s'est aussi présentée comme une création du pouvoir discursif, comme un édifice dans lequel, pour les premiers géomètres, le verbe et la pensée géométrique ne formaient qu'un. Le philosophe est celui qui entend ressaisir tout le donné par le verbe pour en dégager, pour en exprimer, encore une fois par le verbe, ce qui en fait la valeur. Le verbe représente d'ailleurs le seul moyen dont il ne sera jamais privé, le verbe dont la substance est constamment renouvelée par l'inépuisable spontanéité du pouvoir discursif.

Nous disons spontanéité, nous ne disons pas exercice dans l'arbitraire ou création au hasard. Le verbe se plie aux conditions et se prête aux intentions de son emploi. Mais il le fait comme un organisme capable d'adaptation dans la persistance de son existence propre. Il est, à chaque instant, ce que son histoire fait de lui, ce qu'a fait de lui, en particulier, l'histoire de la connaissance, à l'expression de laquelle il a participé. Disons-le d'un mot, le verbe est toujours en situation de connaissance. Disons-le de façon encore plus explicite: un langage reflète la situation de connaissance commune à ceux qui l'emploient en commun.

Il n'en est pas autrement du langage du philosophe, des langages des philosophes. Et c'est là ce qui pose des limites à l'extension de sa validité: celle-ci n'est pas intemporelle, inconditionnelle, elle est simplement celle d'un langage engagé dans une histoire.

Certes, la visée du philosophe ne s'arrête pas devant la science qui ne peut être pour elle que l'une des activités dont l'homme est capable. Il lui faut, à sa façon, c.-à-d. discursivement, l'intégrer à son univers. Il doit l'envelopper, en quelque sorte, d'une trame discursive pour s'en saisir comme il prétend le faire de tout ce qui est essentiel. Mais le filet qu'il jette ainsi, aura-t-il à la fois la force et la finesse indispensables pour retenir tout ce qui donne son poids et son sens à la recherche scientifique?

Il faut, en effet, tenir compte du fait que l'activité discursive ne s'exerce pas sur un plan seulement, qu'elle ne se situe pas à un niveau seulement, qu'elle ne connaît pas seulement une unique forme d'expression qui serait celle du philosophe. La géométrie pure, nous l'avons déjà dit, est aussi un édifice discursif. Sous l'un de ses aspects, c'est aussi une trame de mots qu'on jette sur ce qu'elle doit saisir. La géométrie a d'autres aspects, mais la philosophie n'en a-t-elle pas aussi? Serait-elle plus que l'ombre d'elle-même si elle ne désignait pas, à travers sa discursivité, une réalité émergeant irréductiblement de celle-ci? Ce qui vient d'être dit pour la géométrie est aussi valable pour l'ensemble des mathématiques. Celles-ci constituent aussi un discours susceptible d'être déployé selon les lois de sa propre cohérence. Ceux qui en ont suivi, ne fût-ce que le long d'une courte vie, le développement et le progrès savent que les mathématiques sont actuellement le produit le plus pur et le plus rigoureux du pouvoir discursif dont tout discours ordonné donne une preuve d'existence. Objectera-t-on que le mathématicien ajoute des signes aux mots qu'il emploie? Ce serait lui faire un reproche de ce qui fait précisément son pouvoir de création. Mais, d'autre part, n'arrive-t-il jamais que le philosophe en fasse autant? Sous l'angle de l'intention de discursivité, le langage courant, le discours des philosophes et le discours des mathématiciens vont dans le même sens: ils tendent tout droit à "mettre en forme discursive," pour un temps ou pour tous les temps, le rapport de *ce* dont ils prennent conscience d'eux-mêmes et de *ce* dont ils prennent conscience des autres et du monde. Mais ils ne réalisent pas cette intention au même degré, au même niveau. Le discours mathématique est celui par lequel le pouvoir discursif trouve sa *forme d'intervention* la plus stable et la plus rigoureuse. Sur ce fond, la figure que dessine un physicien tel qu'Eddington est comparable à celle d'un philosophe pour lequel le langage est le moyen d'appeler à une certaine forme d'existence, quelque chose qui n'accèderait pas sans lui à ce degré, à ce niveau d'existence. Le physicien qui cherche à mettre en

évidence la structure mathématique du monde des phénomènes
(d'autres diront qu'il cherche à lui imprimer une telle structure)
ne fait pas autre chose que ce que le philosophe entend faire: il
jette sur tout ce qui vient à lui la trame indéchirable d'un réseau
déductivement cohérent. C'est là sa façon de comprendre, sa façon
d'expliquer, sa façon d'exprimer. C'est là sa façon de se saisir de
ce qu'il rencontre ou de ce qui le rencontre, à la fois par la pensée
et pour l'action. Il se fait ainsi, nous donnons à dessein un tour
particulièrement aigu à cette façon de dire, l'agent du pouvoir
discursif.

Mais il ne le fait pas dans des conditions quelconques, il le fait
dans une situation précisée que l'activité discursive n'a pas été
la seule à dégager. Le physicien qui entend énoncer le monde en
termes de mathématiques, n'a pas la capacité de le faire, de façon
univoque et, par conséquent, nécessaire. Il n'est pas le seul à
avoir voix au chapitre. Les structures mathématiques qu'il érige
en systèmes explicatifs et descriptifs appartiennent à un horizon
d'énonciation qui ne comporte pas, par lui-même, le critère de
leur justesse. Car le critère de leur cohérence mathématique ne
suffit pas, c'est là la grande leçon que la poursuite de l'intention
discursive, à travers l'expérience du physicien, a fini par dégager
et imposer comme un fait sans réplique. Le physicien qui énonce
ou qui cherche à le faire, doit avoir les yeux fixés sur le physicien
qui expérimente. L'horizon d'énonciation du premier tient à l'horizon d'expérimentation du second comme la pensée tient au
geste. Or, dans le cas particulier du physicien ce que le geste découvre n'est pas d'avance dans la pensée. Bien plus, ce que le
geste découvre n'est, le plus souvent, qu'un ensemble d'indices
que la pensée suit comme des traces, mais dont elle ne fait pas sa
substance. En d'autres termes, ni le matériau ni l'objet du discours énonciateur ne sont donnés au préalable, de telle façon que
l'un puisse être simplement mis en place comme expression juste
de l'autre. Il arrive, au contraire que, se déployant par l'intermé-
diaire du discours mathématique, le pouvoir discursif ait à dis-
tinguer et à choisir à la fois, en un seul acte décisif, ce qui accèdera
à l'existence énoncée et des moyens par lesquels cette existence
lui sera conférée.

Il arrive que le physicien s'acquitte de cette tâche de la façon
la plus naturelle, sa familiarité avec le double aspect de son enga-

gement lui faisant, en quelque sorte, une seconde nature. Mais il arrive aussi qu'il prenne conscience de toute la problématique de ce double engagement. Il rencontre alors, pour son propre compte, toutes les questions que le philosophe avait rencontrées avant lui, questions qui s'évoquent par les mots de connaissance, réalité, expérience, fait, idée, concept, principe, loi, etc. Mais le niveau où il les rencontre n'est pas le même que celui où le philosophe se trouvait placé devant les péripéties les plus décisives de l'expérience du physicien dans lesquelles il se trouve encore placé s'il n'a pas pris conscience de ces dernières. Ainsi, chose paradoxale, au sein même du domaine que le philosophe croyait avoir enveloppé de sa saisie discursive, tous les problèmes qu'il se posait (tout au moins ceux qui touchent à la connaissance du monde réel et à notre existence propre au sein de ce monde) se reposent à nouveau, mais dans une situation nouvelle, avec des exigences aggravées, dans des conditions de complexité et d'acuité encore inédites. Le physicien qui ne se contente pas d'être ce qu'il est avec toute la sécurité et toute l'assurance que lui confèrent sa place et sa fonction dans une tradition scientifique déjà éprouvée, mais qui se préoccupe de sa situation et cherche à s'en faire une idée aussi claire et aussi cohérente que possible, celui qui (sans toujours le dire) entend en faire l'objet d'une nouvelle saisie discursive n'est pas autre chose qu'une incarnation moderne du philosophe, au sens le plus immédiat et le plus entier de ce mot. Il est engagé à la fois dans l'effort pour connaître et dans la réflexion sur cet effort: l'authenticité de sa prise de position ne peut qu'en être renforcée. Pourquoi la personne et l'oeuvre d'Eddington méritent-elles notre admiration et notre attention? C'est qu'en lui le savant ne se sépare pas de ce philosophe de la connaissance moderne, mais que l'un et l'autre ne prennent tout leur poids que par le fait d'être indissolublement alliés.

Je pense avoir dit par là-même tout le prix que j'attache à l'étude de M. Yolton et tout le plaisir que j'ai eu à contribuer à la couronner.

F. Gonseth
*Zürich*

Eddington was not unique among scientists who have been interested in the philosophical implications of the presuppositions and techniques of modern science, nor was he unique in being condemned by philosophical critics for writing bad philosophy; but he was perhaps a rare example of a scientist who was highly esteemed by his contemporaries for his scientific discoveries and ability and one whose reflections upon these discoveries were condemned thoroughly by both scientists and philosophers. He commanded sufficient prestige to be elected Gifford lecturer in the year 1927, and in 1937 he was further invited to give the Tarner lectures at Cambridge. The publication of the Gifford lectures was popularly received. One of Eddington's characteristics which brought him this popularity was undoubtedly his willingness to discuss freely all sorts of possible implications of modern physics, a willingness to depart from the rigid techniques of his own field and to offer speculations upon their broad philosophical significance. His two works published prior to his Gifford lectures, *The Mathematical Theory of Relativity* and *Space, Time, and Gravitation*, were more technical books than *The Nature of the Physical World*, but they contained the germs of most of his later thinking. It was not until the publication of his Gifford lectures that Eddington lost some of the prestige he had acquired as a scientific thinker among scientists, for in that work he extended the many philosophical implications of physics which had until then comprised only a small portion of his earlier works. After these lectures, he was definitely launched upon the path of speculation which brought him into conflict with many of the men in his own field. But it is interesting to note that he rounded off his series of publications by a return to a rigid formalism in the posthumous *Fundamental Theory*, a formalism which sought to sanctify the most radical of his philosophical interpretations of modern science by mathematical christening. Mathematics was his point of departure and his resting place. No appreciation of his philosophy of physics can be made without being fully aware of the dominant role played by mathematics in his thought. It served him not only as a tool for formulating the data of science into a deductive and coherent system, but as the background for some of his more radical interpretations of science, such as his claim for *a priori* structural knowledge. But unlike many other scientists who have philosophized about science, Eddington cannot be properly evaluated until we have placed him in his historical context. His philosophy of science was an exercise in epistemology, and his epistemology was essentially and uniquely British. The analysis of the knowledge situation made by Russell in his various books strongly influenced Eddington. Many of the doctrines formulated and accepted by Russell reappear in Eddington, but these doctrines themselves have a long historical past ranging from Russell, Price, Broad and Moore to Locke, Berkeley, and Hume. The methodology and the ontology of this tradition have been preserved with few changes from Locke's day to Eddington's. Although Eddington's discussion and development of these various doctrines of the British tradition is not always marked by clarity and precision, many of the traditional problems, such as the concept of reality, the relation between pure empiricism and rationalism, are given an analysis which brings out the significance of these problems better than is done by Eddington's philosophical contemporaries. There are confusions and ambiguities and even misguided speculations in all of Eddington's works. These make for tedious reading and easy condemnation. But when placed in their historical perspective, many of the confusions can be clarified and given their proper place in the context of dualistic theories of knowledge and reality. I would not claim that Eddington's philosophy of science by itself is valuable reading or that it should be the subject of extended studies by those interested in the subject. But to those who have an interest in the British tradition of epistemology as well as to those who are curious about the many controversial claims of his philosophy of science, a close analysis of Eddington's system in this historical context is worthwhile.

This study is a revised version of an essay submitted to the

Institut International des Sciences Théoriques at Brussels for the Eddington contest it sponsored in 1951. This essay was awarded first prize by the Institut in January 1956. I wish to thank the members of the Institut, in particular Father Dockx, for their interest in this study and for their kindness and hospitality in Brussels when the award was made. I am especially indebted to Professor F. Gonseth for his encouragement to publish this study and for his continued interest in it.

Wherever there are quotations in the text from books whose titles are in French in the bibliography, the translations are my own, unless otherwise indicated. For convenience of reference throughout, I have adopted the following symbols for those works most frequently used:

BJPS  – *British Journal for the Philosophy of Science*
MTR   – *The Mathematical Theory of Relativity*
NPS   – *New Pathways in Science*
NPW   – *The Nature of the Physical World*
PPS   – *The Philosophy of Physical Science*
STG   – *Space, Time, and Gravitation*

# CONTENTS

"The ideal which this philosophy strives after is a mathematical world-formula, by which, if all the collocations and motions at a given moment were known, it would be possible to reckon those of any wished-for future moment, by simply considering the necessary geometrical, arithmetical, and logical implications. Once we have the world in this bare shape, we can fling our net of *a priori* relations over all its terms, and pass from one of its phases to another by inward thought-necessity."

WILLIAM JAMES, *The Principles of Psychology*, vol. II, p. 667.

# CHAPTER I

# THE OPERATIONAL ATTITUDE

Eddington was one of the first scientists who appreciated the revolutionary changes in our philosophical view of the world entailed by the new relativity theories in physics. He was concerned to make explicit the philosophical foundations of the new science, but he had no pretensions as to the adequacy of these foundations for a complete understanding of our world. Science and its philosophical presuppositions provided only one of several approaches to reality. Eddington was careful to indicate the precise bounds of the physical sciences, demarcating them from religion and esthetics and all other normative fields of thought. "Life would be stunted and narrow," he wrote in his Gifford lectures, "if we could feel no significance in the world around us beyond that which can be weighed and measured with the tools of the physicists or described by the metrical symbols of the mathematician." (NPW, p. 317) Later, in his Tarner lectures, while stressing the limitations of scientific knowledge to observation, he shows the same recognition of the partial view of the sciences towards the world.

We do not deny that knowledge which is not of an observational nature may exist, e.g. the theory of numbers in pure mathematics, and noncommitally we may allow the possibility of other forms of insight of the human mind into a world outside itself. But such knowledge is beyond the borders of physical science, and therefore does not enter into the description of the world introduced in the formulation of physical knowledge. (PPS, p. 10)

In these lectures, he was concerned with developing the epistemology implicit in physical knowledge; he was not concerned to sketch other possibilities of different forms of knowledge. In his Gifford lectures, where he was dealing with the wider setting of physical knowledge, Eddington embarks on various speculative problems traditionally associated with the philosophy of science but which are really outside the scope of physical knowledge. He shows a tendency to draw non-scientific conclusions from the data of science. In one passage in *The Philosophy of Physical Science*, he suggests that the "endeavour of the scientific philosopher must be to extend this rational correlation of experience from a limited field of experience to the whole of experience" (p. 187), but this remark is meant only to apply to the philosophy of science being developed in that treatise. Most of his discussions of non-scientific problems, such as the mental character of reality, the nature of God, or the possibility of human freedom, are offered more as possibilities still consistent with the data of science than as implications of those data. These are peripheral problems upon which Eddington was not averse to passing judgment, but they carry him outside the scope not only of physics, but of the philosophy of the exact sciences as he himself conceived that task.

The philosophy of the physical sciences for him consisted in explicating the epistemological and methodological presuppositions of the exact sciences. Methodologically, the physical sciences are concerned with establishing correlations between those experiences which can be weighed, measured or otherwise manipulated by the tools of science. The aim of such correlations is "to disclose the scheme of the recurrencies in the combined experience of conscious beings." (NPS, p. 45) Constancy, repetition of patterns in sensory experience, becomes the stepping stone from unscientific to scientific awareness. It is the recurrent features of experience which science singles out for special treatment, and it is these features which science seeks to correlate internally and with the rest of experience. The basic objective is to analyze experience into irreducible components, where simplicity is the guide for reduction. We shall find Eddington suggesting that simplicity is a fundamental category of human thought, actually determining the character of our scientific data; but whether we finally agree with him in this contention, it is clear that he himself was driven by a desire for simplicity in his philosophical interpretation of the physical sciences. Simplicity is taken as a demand for reduction.

As applied to the actual operations of physical science, Eddington contends that the scientific observer need have only one eye and none of the other normal sense modalities, since the necessary data for scientific knowledge are comprised solely of pointer readings. "The eye need not have the power of measuring or graduating light and shade; I think it is sufficient if it can just discriminate two shades so as to detect whether an opaque object is in a certain position or not." (NPS, p. 12) Such a reduction of our sensations makes it possible for science to concentrate upon visual sensations alone.

> It is sufficient if we can distinguish by direct awareness a particular class of sensations, which by itself is sufficient to reveal all that is known of the physical universe. Ideally all our knowledge of the physical universe could have been reached by visual sensation alone – in fact, by the simplest form of visual sensation, colourless and non-stereoscopec. We can therefore regard an item of physical knowledge as an assertion of what has been or would be perceived visually. (PPS, p. 197)

Visual sensations of pointer readings have become the data of the exact, mathematical sciences. All observations become, as Whittaker has remarked, topographical relations, or more precisely, "the rules governing these topographical relations."[1] It is important to recognize the connection between Eddington's reduction of the ordinary world to pointer-readings and the mathematicization and geometricization of science, for this reduction applies only to the exact sciences, to those sciences or those parts of science which have proven themselves amenable to mathematical formulation. The exact sciences are only a section of science in general and the conclusions and theories operating in them may not hold true for science as a whole. Such a separation is especially necessary in view of the very abstract character of knowledge typical of exact science. Some of the paradoxical and dubious features of Eddington's philosophy are clarified when we bear this distinction in mind, for science, when extended to include psychology, biology and the social sciences, need not so reduce and constrict the ordinary world. It is only when the formalizations of the exact sciences have been copied by the other sciences that these sciences introduce the same kind of concentration upon methodology which has characterized the formalizations of physics. The philosophical interpretation of science, whether directed to physical or social science, has been characterized in the twentieth century by a dominant stress upon such methodological considerations. One aspect of the positivistic philosophy has been concerned with formalizations of

[1] "Eddington's Theory of Constants," *Mathematical Gazette*, vol. 29, p. 142.

concepts employed in science or, more recently, with semantical operations in general. The more traditional concerns of the philosopher have become lost through this double orientation toward science and formalization. Until recently, it was the fashion to claim that traditional philosophy had been naively misled into thinking its problems were genuine. The reductionism inherent in positivism affected not only man's world but man himself. Such a reductionism within philosophy can only be defended by a similar redefinition of the nature of the philosophic enterprise. But if we are concerned only with the necessary tools of the physical sciences, it is clear that Eddington's reduction of man follows from the introduction into science of exact mathematical measurements.

> If we exclude the process of pure counting, by which, for instance, we find the number of Jupiter's moons, these conventional processes [of measurements] are found to consist of visual observations of the alignment of two marks, together with counting. In *exact* science these are the only observations that are necessary.[1]

In other words, in the exact sciences everything which cannot be symbolized by a number introduced by the conventional means of measurement is omitted. The distance from and even distortion of the common sense world by the exact scientific world picture is brought about by this reduction of the complexity of the former to the enforced mathematical simplicity of the latter. The poetry of existence "fades out of the problem," as Eddington admits, and we are left with only the pointer readings taken by a contemporary cyclops. The fullness of the familiar observer's perceptual experience is replaced by the monotonous repetition of data from calipers, meters, rods, etc.

Eddington was careful to specify that these pointer readings must be observed in context, that while an isolated reading would constitute an observational datum, nevertheless it is not with such segregated readings that science is concerned. "For scientific knowledge the association with other pointer readings is an essential condition; and we may therefore describe physical knowledge as a knowledge of the associations of pointer readings."

---

[1] "A New Treatment of the Theory of Dimensions," G. B. Brown, in *Proc. Physical Soc.*, vol. 53, p. 419. See also his letter in connection with the Dingle-Jeans-Eddington controversy in *Nature*, vol. 48, October 25, 1941, p. 504.

(PPS, p. 100) The context of pointer readings links together to form an interconnected system.

> When for example, we determine the intensity of a magnetic field, we associate with it the time and the co-ordinates in space of the point to which the determination applies. The magnetic intensity is then the primary pointer reading, and the co-ordinates in space and time are the secondary pointer readings. But the chain of pointer readings does not stop here. Tertiary pointer readings are required to identify the system of co-ordinates used, and to determine its metric; but these tertiary readings are common to all items of knowledge referred to the co-ordinate frame, and (unlike the secondary pointer readings) are not determined afresh for every primary reading. (*Ibid.*)

The same contextual emphasis is present in Eddington's discussion of the analytical procedures of science where he describes analysis as involving "a whole as divisible into parts, such that the co-existence of the parts constitutes the existence of the whole." (*Ibid.*, p. 118) He was fully aware also that the efficacy of science demands the cooperation and interconnection of many observers. Like simplicity, contextual relations are asserted to be a basic category of thought determining the form of our scientific knowledge. Whether we agree with the primacy of this alleged category as with the question concerning simplicity, is irrelevant for a general understanding of Eddington's view of scientific method or of the nature of scientific knowledge. For he can hardly be disputed when he says "an intellectual activity begins when we relate our perceptions to one another." (*Ibid.*, p. 114) Awareness itself demands comparison and diversity. Out of such comparison arises the kind of interconnected knowledge characteristic of science. "Knowledge of the relatedness of sensory perception, e.g. the sound of thunder following the flash of lighting, is the beginning of science." (*Ibid.*) But science moves quickly from such common perceptual phenomena to the constructions and inferences, abstractions and symbolisms of its more advanced stages. The context of pointer readings does not include the ordinary perceptual world since such readings are not made by the ordinary man. The stress upon context for Eddington is an emphasis upon system, a system of pointer readings. But what gives the system as a whole its meaning and significance? Does not the context have to be expanded in the direction of ordinary perceptual experience before the readings can be interpreted?

Köhler has urged that even in the exact sciences, something more than pointer readings taken in context with other pointer readings is required before their abstract picture of the world can be constructed.

> Pointer-readings as such are the same perceptual data whatever we intend to measure. What we need are *different* perceptual contexts inside of which pointer-readings acquire in each case a particular meaning, 'current' now, 'pressure' another time, and so on.[1]

Köhler's point is that in and by themselves pointer readings could not supply these meanings. Some reference to what he calls phenomenal experiences must be made in order to distinguish a 'current' from a 'pressure' or 'resistant' situation. Köhler's argument on this point is directed by his desire to tie all scientific knowledge to the common sense world; but even though we do not wish to sanction this strict dependence of science upon common sense, Köhler is correct in suggesting that some basis is presupposed in Eddington's account of the data of the exact sciences for giving meaning to and distinguishing between pointer readings.

The operationalists insist that this expansion of the context of data need include no more than the operations used in obtaining the data; if these operations are, as they frequently are in physics, abstract and truncated when viewed from the perspective of our common perceptual experiences, the context of science will be similarly reduced for the operationalist. Köhler as a psychologist does not wish to sever perceptual from scientific experience in this way. But whatever the nature of the scientific context, it is clear that the operations are as important as the data. Observation and manipulation must be correlated with the marks on the gauges. Eddington did not wish to maintain that every item which has been accepted by science as a datum has actually been certified by the court of observation, since many aspects of science rest upon data not directly observed or taken from a set of pointer readings; but the basic presupposition of science is that all such data could be verified by observation. Every item of physical knowledge "must be such that we can specify (although it may be impracticable to carry out) an observational procedure which would decide whether it is true or not." (PPS, p. 9) For

---

[1] *The Place of Value in a World of Fact*, p. 159.

this reason, he preferred to call the knowledge of the physical sciences hypothetical-observational. The amount of actual observation which goes into physical knowledge is very small. When for example, "in the process of reducing observations a 'correction' is applied, observational knowledge of an actual experiment is replaced by hypothetical-observational knowledge of what would have been the result of an experiment under more ideal conditions." (*Ibid.*, p. 12) The gain achieved by such hypothetical substitution results in a systematized knowledge which can be "gathered into a coherent whole, whereas actual observational knowledge is sporadic and desultory." (*Ibid.*, p. 13)

If the knowledge yielded by the exact sciences is hypothetical-observational, the basic terms in the linguistic formulation of those sciences must be similarly defined. The physical quantities of length, angle, velocity, force, potential current are themselves defined "according to the way in which we actually recognize them when confronted with them."(NPW, p. 254) The function of the operational attitude in Eddington's general system will be seen to reduce to a linguistic formulation of one large section of the data of science. But like many others who have accepted the operational theory of meaning, Eddington frequently argues as if this theory were not only a semantic tool but a theory of reality as well. The sphere of the meaningful is restricted by the strict operational method of definition. Entities which cannot be observed must be excluded.

When an unobservable is introduced into a statement which professes to be an expression of physical knowledge, the statement is usually rendered meaningless; as an item of physical knowledge it must assert the result of a specified observational procedure, and the intrusion of a term which is without observational meaning causes a hiatus in the specification. (PPS, p. 39)

The methodological insistence upon operational contexts of meaning carries its ontological implication. Reality as well as meaning is operationally defined. Phenomenalism replaces realism. The qualities which we formerly ascribed to an external world are now no longer properties of such a realist world.

*Length* and *duration* are not things inherent in the external world; they are relations of things in the external world to some specified observer. ... When the rod in the Michelson-Morley experiment is turned through a right angle it contracts; that naturally gives the impression that something has happened to the rod itself. Nothing whatever has happened to the rod – the object in the external world. Its length has altered, but length is not an intrinsic property of the rod, since it is quite indeterminate until some observer is specified. (STG, p. 34)

As he remarked later, "Einstein's theory makes a clear sweep of these pious opinions, and insists that each physical quantity should be defined as the result of certain operations of measurement and calculation." (NPW, p. 255) Bridgman has recently expressed this position more succinctly. "The fundamental idea back of an operational analysis is simple enough; namely that we do not know the meaning of a concept unless we can specify the operations which were used by us or our neighbour in applying the concept in any concrete situation."[1] Eddington suggests that we can think of these terms as "something of inscrutable nature to which the pointer reading has a kind of relevance," but he points out that such a belief lies outside the scope of the exact sciences and would only hinder its progress. (NPW, p. 255) He had written in *The Mathematical Theory of Relativity*: "but in an experimental science we have to discover properties not to assign them; physical quantities are defined primarily according to the way in which we recognize them when confronted by them in an observation of the world around us." (p. 1) A strict operationalism would dispense with talk of the world around us, replacing 'observation' by 'operation.' Eddington's formulation of the operational attitude was not always strict; but in the operationalist passages it is clear that the drive for simplicity leads him to agree with the dicta of Bridgman by excluding all entities from the scientific language which cannot be subjected to rigorous confirmation by careful and controlled operations. So exact have the physical sciences become that they have found it necessary to admit as meaningful only those statements and terms which conform to this general demand of observational and operational control.

In such a system of meaning, length and time play a primary role "because in general the definitions of other physical quantities presuppose that length and time-extension have been defined, and any ambiguity of their meaning would spread through the whole superstructure." (PPS, p. 73) The definition of all measure-

---

[1] "The Nature of Some of our Physical Concepts," in BJPS, vol. I, p. 257.

ments in terms of these two basic measures is again the result of the rigorous reduction of operations and premises characteristic of the exact sciences. The goal of these sciences has been to strive for an orderly deductive system, a goal which, in their exaggerated formulations of it, has exposed both Eddington and Milne to violent criticisms from their contemporaries as a result of their claim to have thoroughly completed such a compact deductive system. Not only have all measurements been reduced to two primary ones, but these men have claimed to be able to derive most of modern physics from a small number of premises. Simplicity and systematization constantly function for Eddington as guide posts leading to the reduction of the number of concepts employed by science and to the ordering of these concepts into a deductive pattern. Within his operational theory of meaning, it is simplicity which plays the dominant role; length and time becoming the primitive concepts.

> The dimensions of a physical quantity consist of one or both of the symbols for the two primary operations of measurement of length and time, together with the respective indices representing the powers to which the number so obtained are raised in order to conform to the definition of the quantity.[1]

The usual stress is upon length itself; concentration upon length reflects a latent desire to absorb the measurement of temporal duration into a measure of length.

> A definition of length which specifies a way of determining lenghts observationally is indeed the most urgent requirement of all; for when we come to examine what is actually measured in any kind of experiment, it is nearly always a length or spatial measure – the length of a thread of mercury in a thermometer, the shift of a bright spot on a galvanometer scale, the dissplacement of a dark line in a spectogram, etc. (PPS, p. 71)

In the Prologue to *Space, Time, and Gravitation*, Eddington discusses the way in which Einstein's relativity theory has rendered necessary this revision of the meanings of terms from the old absolute Newtonian sense to the operational meanings now being applied, stressing again the importance of length as a primary concept. There, the traditional physicist is engaged in conversation with the mathematician and the modern relativist. The point challenged by the latter is the adequacy of Euclidean

[1] G. B. Brown, *op. cit.*, p. 423.

geometry for purposes of the exact sciences. Admitting that length and distance cannot be given a meaning apart from "a quantity arrived at by measurement with material or optical appliances," the physicist seeks to retain the older view of length as an absolute quantity, insisting at the same time that it has to be measured by a rigid scale. The physicist of this dialogue is presented as aware of the difficulties attached to relinquishing the old conceptions. Eddington later expressed these difficulties in his *The Mathematical Theory of Relativity*, clearly outlining the main features of the operational attitude and its difference from the older view.

> The physical quantity so discovered [through measurement] is primarily the result of the operations and calculations; it is, so to speak, *a manufactured article* – manufactured by our operations. But the physicist is not generally content to believe that the quantity he arrives at is something whose nature is inseparable from the kind of operations which led to it; he has an idea that if he could become a god contemplating the external world, he would see his manufactured physical quantity forming a distinct feature of the picture. By finding that he can lay $x$ unit measuring rods in a line between two points, he has manufactured the quantity $x$ which he calls the distance between the points; but he believes that that distance $x$ is something already existing in the picture of the world. (MTR, p. 1)

The prerequisite for this older belief is that the measuring rod is rigid and does not alter. But the relativist of the Prologue to *Space, Time and Gravitation* points out that we could never devise a test to determine whether our scale was rigid or not, for such a test would entail an infinite series of testing scales. In fact, he argues, it is a matter of definition that the rigid rod cannot change length for, "if a metre is defined as the length of a certain bar, that bar can never be anything but a metre long; and if we assert that this bar changes length, it is clear that we must have changed our minds as to the definition of length." (p. 4) The relativist recognizes some meaning to the phrase, "correcting for defects in the measuring rod," such as those caused by heat. "Ordinary scales have defects – flexure, expansion with temperature, etc. – which can be reduced by suitable precautions; and the limit to which you approach as you reduce them, is your rigid scale." (p. 4, 5) On the other hand, the supposed distortion caused by being in the field of a strong magnet is of a different nature.

> You correct measures, when they are untrue to standard. Thus you correct the readings of a hydrogen thermometer to obtain the readings of

a perfect gas-thermometer, because the hydrogen molecules have a finite size, and exert special attractions on one another, and you prefer to take as standard an ideal gas with infinitely small molecules. But in the present case, what is the standard you are aiming at when you propose to correct measures with the rigid rod? (p. 6)

In this case, we are not dealing with ordinary cases of defects in material, since all materials would be affected by the field of force generated by a magnet. A similar situation arises when the measuring rod is subjected to rapid motion. Thus the suggestion is that in these cases the traditional physicist's position is untenable; since no meaning can be given to 'correction' in these contexts, there is no standard to which we can appeal. An operational definition of 'defect' or 'correction' must be constructed.

You can define these defects without appealing to any extraneous definition of length; for example, if you have two rods of the same material whose extremities are just in contact with one another, and when one of them is heated the extremities no longer can be adjusted to coincide, then the material has a temperature-coefficient of expansion. (p. 5)

The rod with the lowest coefficient of expansion can be called the ideal rigid rod.

In *The Nature of the Physical World*, Eddington sought to elaborate a distinction between essential and casual characteristics as they function in measurement – temperature, strain, and corrosion being labeled 'casual' factors which we try to eliminate in our selection of the material, while the influences of a magnetic field are considered essential characteristics. He does not offer this as a rigid distinction, for he says that both types of factors are themselves determined in relation to the aims of the operational task.

The distinction between casual and essential influences – those to be eliminated and those to be left in – depends on the intention of the measurements. The measuring rod is intended for surveying space, and the essential characteristic of space is 'metric'. It would be absurd to correct the readings of our scale to the values they would have had if the space had some other metric. The region of the world to which the metric refers may also contain an electric field; this will be regarded as a casual characteristic since the measuring rod is not intended for surveying electric fields. (NPW, p. 142.)

The context for science is thus broadened to include not only the operations as well as the data but the intentions, aims, and preferences motivating the use of one technique or instrument over another. Such an expansion leads in the direction of ordinary perceptual and psychological experiences and may therefore be questionable within the operational approach. It is clear that the talk of influences, whether they be casual or essential, is not permissable in a physics defined completely in operational terms, since such talk presupposes our measuring rod to be possessed of properties which preexist our operations.

> We should begin with the operation and its result and then, if we wish to speak of a property (which I do not think we shall do), define it in terms of that. The result of each operation should initially have its own name, and if we find that different operations yield approximately the same result, then we have made the empirical discovery that the quantities represented by the different names are approximately equal.[1]

Instead of saying that heat alters the length of the rod, a consistent and thorough operationalism would demand some such operational definition of 'defect' as offered by the relativist in Eddington's Prologue. But what is it that leads us to seek for a rod with a low degree of coefficient of expansion? The operational definition of 'defect' is obviously at the same time a definition of 'coefficient of expansion', but can we state in operational terms the reasons for desiring a rod of this nature?

The answer to this question is bound up with the talk about defects and correction for errors, since even though Eddington was able to define 'defects' operationally, the initial problem has only been relocated. The operational definition of 'defects' makes a comparison between two rods, but such comparisons implicitly contain a preference for one rod over another. This preference now contains the bothersome ingredient. If Bridgman is to be followed when he argues that the "operational analysis is applicable not only to the meaning of terms or concepts, but to other matters of meaning, as for example, to the meaning of questions", such that "I do not know what I mean by a question until I can picture to myself what I would do to check the correctness of an answer,"[2] then either we can find an operational answer to the question of the basis of the preference for a rod with a low degree of coefficient of expansion, or it must be admitted that the operational attitude itself rests upon and presupposes a non-opera-

---

[1] Dingle, H. "A Theory of Measurement," in BJPS, vol. I, p. 7.
[2] Bridgman, *Nature of Physical Theory*, p. 11.

tional field. Preferences occur at other junctures in the methodology of the sciences, so this is not a trivial issue. Dingle argues
that the "reason why we make some measurements and not
others can be understood only in terms of the ultimate object
of science, which is to find relations between the elements of
our experience."[1] Calling attention to the fact that in the methodology of measurement, some aspects of experience are essential
and relevant while others are unessential and can be and normally are ignored, Dingle raises the question, 'what determines
which details are essential?'. Dingle denies that the answer
is "that we specify all the details that are found to affect the
result and ignore all the others."[2] The real factor is, he insists, "the same as that which decides what measurements we
shall make – namely, that we specify only those conditions that
are necessary to give us results that stand in simple relations to
others."[2] Ignoring the ambiguity of the phrase 'simple relations,'
can we say similarly that the explanation of the preferences for
measuring-rods of low degrees of coefficient of expansion lies in
the fact that with this choice we can establish more simple
relations between experiences than otherwise? Is the desire for
simplicity at work even here? I think not. The real answer for
this preference seems to me to be historical. The discovery that
some measuring rods were preferable to others arose while scientists were still working under the older point of view, where they
considered their measurements as discovering real properties in
nature. Since these supposed real properties were also assumed to
be unchanging, it was a natural consequence to look for measuring rods that did not themselves alter with each new operation
or under certain climatic conditions. This assumption of a real
external world waiting to be discovered underlies many of the
procedures and preferences of the physical sciences. The crucial
question which the operationalist must face is not alone 'how to
exorcise the old demon of common sense realism?', but much
more importantly, 'is it possible to make operationalism as
extensive as Dingle and Bridgman wish?'. It is a retrospective
problem with respect to the measuring rods, the problem of

---

[1] Dingle, *op. cit.*, pp. 11–12.
[2] *Ibid.*, p. 13.
[3] *Ibid.*

whether it is possible to explain the whole of present-day science by reducing to the operational level of meaning. Even the avowed operationalists find it difficult to talk the operational language at all times. In his *Logic of Modern Physics*, Bridgman talks repeatedly of the *accuracy* of measurements and of the *approximate* character of all measurements. Even Dingle, the writer who has argued for the most extensive form of operationalism, is not entirely free from lapses. For example,

> If we take such a simple operation as that of measuring length ... we shall have to describe how to construct a standard measuring rod, including the testing of its material for various characteristics; we shall then have to describe how copies are to be made; and finally, how they are used in *finding the length of the object* we wish to measure.[1]

Difficulty of completion should not, of course, argue against the validity of their recommendation for a complete operationalism. The frequent lapses into the older mode of speech would in any event be unavoidable since it is unusual and artificial to talk the operationalist language; but it appears to me that unless we wish to reduce the explanation of the procedures of the sciences to a situation in which it is sheer coincidence that scientists have selected rods of a low degree of coefficient of expansion, we are forced, in our explanation, to admit some non-operational factors. If we examine physics as it is now, its present procedures and concepts, forgetting its historical growth, we may be able to establish an area of operational meaning; but motivations and preferences are a fundamental part of physics which do not seem amenable to an operational explanation.

The difficulty can be avoided by simply concentrating upon the operations, ignoring the reasons and motivations behind them. The strict operationalist would have to follow just this procedure. But he would also have to admit that the operational attitude is concerned only with formulating scientific terms and concepts according to a definite criterion of meaning. Ontology is reduced to epistemology, the physical world is itself defined as that which the tools of science are capable of describing.

> By defining the physical universe and the physical objects which constitute it as the theme of a specified body of knowledge, and not as

---

[1] *Ibid.*, p. 13. My italics.

things possessing a property of existence elusive of definition, we free
the foundations of physics from suspicion of metaphysical contamination.
(PPS, p. 3, cf. pp. 101, 159)

That world so restricted is a metric world, a world whose
existence and dimensions are circumscribed by the limitations
of possible pointer readings. What lies beyond the hypothetical-
observational method of science has no meaning. The operational
attitude is applied not only to the fundamental terms but to the
general concept of physical reality. But this wider application is
operative only within the operational sphere of meaning; only
when Eddington wished to confine his discussion and interpretation of science to the elucidation of the actual processes of measurement or to lend credence to his subjectivist leanings. We shall
see in chapter VI that he wavered between an operational-phenomenalist definition of objective reality and the old Newtonian
absolute conception which he was the first to condemn when
discussing the new changes in meaning introduced by relativity
physics. His conscious departure from operationalism in his
definition of time in *The Nature of the Physical World* is a clear
example of his ambivalence with respect to the two points of
view. The indecision marked by this oscillation was due to many
factors, not the least of which was his incomplete digestion of the
various epistemological strands in British thought; but it was
due also to his intuition (he nowhere expressed this thought explicitly) that the operational attitude does have its limitations,
that the narrow world of the exact sciences has to be augmented
by a larger perspective. Eddington's concern with formulating an
operational position was, like that of Bridgman's and Dingle's,
motivated by his desire to replace the older realist terminology
by one more in harmony with what he termed the 'selective
subjectivism' of modern science. Einstein's revisions had forced
physicists to discard many of the realist terms accepted without
criticism since the time of Newton; but Eddington did not find
that these necessary revisions had to encompass the whole of
science.

Once the world known by the physical sciences is defined as a
world largely conditioned by the techniques which yield its
results, we can understand that "any world that we may con-
template is no longer an independent existence whose nature

demands or determines them, but rather a logical construct, formed and shaped and modified so as to afford a true picture of the relations which the observations exhibit."[1] We can no longer talk indiscriminatly of the properties of bodies in themselves, since almost all the knowledge we have of bodies is bound up with our means of gaining that knowledge. "I know of no revelation, human or divine, that declares what properties a body shall have, that allows specific heat but proscribes sonority, and that gives permission for viscosity if a liquid is moving slowly but withdraws it if the liquid hurries up."[2] Dingle allows this situation of uncertainty as regards our knowledge of any external world to lead him into a very restricted philosophical view of scientific knowledge.[3] He is certainly correct in urging that the measurements we make are determined in terms of the aims we have, in terms of what features of our experience we wish to analyze. All properties resulting from such measurements are on the same footing: none can be ascribed by the scientist (or by the operational philosopher) to a world independent of the observer or of his tools of measurement.

> The situation which has thus presented itself is part of a more general situation, for always, from the point of view of operations, it is fruitless and meaningless to attempt to establish the existence of anything independent of the means by which its existence is established or verified. The two together, object and means of observation or measurement, form an indissoluble union; either without the other is meaningless, at least in the instrumental domain.[4]

The separation of the 'instrumental domain' from other fields, although not made explicit by Bridgman, suggests a recognition of other areas and other interpretations of science. He admits that the operational theory of meaning is by no means "the only aspect of meaning, but it is often the most important single

---

[1] Dingle, *op. cit.*, p. 5.

[2] *Ibid.*, p. 11.

[3] With his other statements concerning the nature of the operational attitude should be placed his description of this position in his obituary on Eddington. He there says that the operationalist view is "that measurements exist first of all in their own right, representing only the operations which yield them, and that our picture of the external world must take the form imposed on it by the necessity of integrating those measurements into a rationally coherent system." (*Proc. Phys. Soc.*, vol. 57, p. 246)

[4] Bridgman, "The Nature of Some of our Physical Concepts," BJPS, vol. I, p. 266.

aspect, particularly in scientific situations."[1] But the operational attitude is at best only one permissible reading of science. Just as the scientist is concerned to interpret and render intelligible man's experience and his world in terms of physical, biological, psychological or sociological principles of meaning, so the interpretation of science itself is a function of definite principles which control the interpretation. The operationalists have tried to argue that only one interpretation is permissible but Eddington constantly employed a realist as well as operationalist principle in dealing with the significance of science. It would be easy to read Eddington's insistence, on the one hand, on a strict operational attitude, and on the other, his equally strong insistence upon talking of world conditions and structural knowledge of a causal world beyond sensation, as contradictory and inconsistent factors of his general position. The truth of the matter is that Eddington attempted to conform to the need for the operational attitude within the confines of the verbalizations of the actual procedures and concepts used in the exact sciences, but was not willing to concede that this necessary restriction was antithetical to a wider perspective when talking about the context in which the exact sciences occur.[2]

I am not concerned in this study to discuss the possibility of a full operational attitude, although I shall point out several ways in which it seems to me to be in need of augmentation. There are difficulties in carrying out such a program as Dingle outlines for operationalism even within the exact sciences, difficulties which may prove insuperable. But the general dictum of this attitude, i.e., restriction of meaning to the actual operations and procedures of the sciences, can be accepted without raising any of these more subtle difficulties. Whether or not a definition of length, for example, can be formulated which does not ultimately presuppose some non-operational factors, the operationalists are correct in arguing, for the purposes of the exact sciences, that the

[1] *Ibid.*, p. 257.

[2] It is quite correct to point out that Eddington's approach to operationalism enjoyed a certain growth and development. In *The Mathematical Theory of Relativity* physical quantities are considered not as the result of operations only but as the relations between objects and measuring rod. (p. 5) By the time of the Gifford lectures, operationalism had a firm hold upon his thought. His statements of the position there could, in most cases, have been taken from Bridgman. (cf. esp., p. 154)

operational theory of meaning (or some close approximation) is a legitimate theory. But we must distinguish carefully between science and philosophy or, within philosophy itself, between those attitudes which are patterned after the methodologies of science (the rigidly empirical) and those broader if less strictly empirical orientations. It has been the effect of the contemporary stress upon making philosophy a science which has sanctioned the operationalist and positivist attitudes. However, it is wrong to force Eddington's philosophy of science into this positivist mould; it can be so forced only by employing the charge of inconsistency liberally throughout his writings. A more effective and significant analysis of Eddington's interpretation of science results from expanding our demands upon a philosophical interpretation of science. In the expanded version, there are two tasks to be undertaken. The first is to offer a verbalization of the concepts of science such that this linguistic product will be consistent with what is actually done in the sciences. The second is to set in their wider philosophical context both the actual procedures of the sciences and the linguistic system thus developed as their expression. For the first task, operationalism is at least one linguistic formulation which expresses fairly well what goes on in the exact sciences. The meanings with which the workers within these fields are concerned are just those meanings which can be extracted from the readings and operations they make. But in setting forth the philosophy of these sciences, we have to be concerned with ontological as well as linguistic matters. The operationalist argues that his theory of meaning reveals also the theory of reality validly inferrable from science. What the operationalists have failed to see is that science by itself has no ontology. An ontology can be read out of physical science only when we are equipped with the necessary interpretative principles. Operationalism can become an ontology only with a philosophy which takes the general phenomenalist dictum (the real is defined in terms of actual and possible experiences) as its ontological criterion. It is even doubtful whether an interpretation of the meaning of scientific terms can be extracted from science without the importation of meaning-rules into science; but the minimum that the operationalist attitude can claim by way of extraction from the operations of science is a language expressing the meanings of those scientific terms necessary for the practice of the scientist. There are several ontologies implicit in science; they are rendered explicit only after we approach science with our ontological convictions.

Neither the outspoken operationalists nor Eddington have ever been very clear about the relation between meaning and ontology, or between ontology and science. But though he saw the attractiveness of operationalism in theory of meaning, Eddington could never really accept it in theory of reality. We find him talking now about the need for a careful operational definition of terms and the restriction of meaning to the actual procedures of the exact sciences, while at other times freely transcending these restrictions. At the very beginning of his final analysis of the philosophy of the physical sciences, he calls attention to the possibility of such a wider context. "To a wider synthesis of knowledge, of which physical knowledge is only a part, we may perhaps correlate a 'world' of which the physical universe is only a partial aspect." (PPS, p. 10) The doctrine of structure and the causal theory of perception developed and accepted in the same book are expressions of this wider synthesis. He was concerned both to offer a verbalization of the exact sciences taken by themselves and to call attention to the philosophical presuppositions lying behind these sciences, presuppositions which take us beyond the operational restrictions. So long as Dingle's injunctions about extending the operational attitude are taken only to cover the talk which goes on within the exact sciences, these injunctions can be accepted. But if they are taken as Dingle meant them to be, as injunctions applying to all our verbalizations and all our concepts about the sciences, then Eddington is on the more secure and satisfactory ground in allowing terms which refer beyond the immediate phenomenal world.

Such an extension of the operational attitude as envisaged by Dingle does not, when properly restricted, necessitate the denial of an objective external world. What Dingle recommends is that the broad philosophical questions of the nature of the physical world – its properties, its relations with observers – are so ambiguous and inessential to science that they can be abandoned. Everything that is desired can be accomplished by adopting the operational-phenomenalist view and many puzzles can be avoided, so long as we offer this view as applying only to the actual procedures of the exact sciences. But Dingle fails to discuss the general context of the exact sciences; he does not recognize the need for exposing the non-operational presuppositions of those sciences. Eddington says at one point that the "continual advance of science is not a mere utilitarian progress; it is progress towards ever purer truth." (NPW, p. 285) Eddington qualified this statement by insisting that the truth "we seek in science is the truth about an external world propounded as the theme of study, and is not bound up with any opinion as to the status of that world." But unless he would settle for a coherence definition of truth, as I do not think he would, the concept of truth, especially of 'purer' truth, entails more than a simple phenomenalism. In *Space, Time and Gravitation*, he makes a similar assertion without any qualification: "I am not satisfied with the view so often expressed that the sole aim of scientific theory is economy of thought. I cannot reject the hope that theory is by slow stages leading us nearer to the truth of things." (p. 29) Earlier he said of relativity theory: "It would be rash to suppose that it reaches finality; but it bears all the indications of being one of the more permanent stages in the advance towards truth."[1] Dingle talks as if the goals of science are only the manipulations of symbols and the resulting measurements. Are we to conclude that the only legitimate aims of science are symbol manipulation and pointer-reading coincidences, and that the suggestions of a wider meaning found in Eddington's works must be discarded as vain speculation? Or can we go beyond Dingle's restricted description while remaining faithful to the operational-phenomenalist attitude? We shall find that Eddington met this problem many times and tried to make his operationalism consistent with his belief in a broader philosophical base for science. It is important to understand just what are the demands of the operational attitude, to see how it applies to the microscopic and macroscopic realms of science, and to grasp the relation between the operational concept of reality and the realist concept said to be ingredient in common sense. For positivism, the philosophical counterpart of operationalism in science, has yielded only the most barren results, failing to afford a foothold for our more fertile concepts

---

[1] "The Philosophical Aspect of the Theory of Relativity," *Mind*, October 1920, p. 416.

of reality. I shall attempt to show in this study that Eddington's operationalism was clearly meant to be only one side of his philosophy of science, and that the dualism which constituted the ontological setting for his philosophy is a legitimate if difficult interpretation of the physical sciences.

# CHAPTER II

# THE CAUSAL THEORY OF PERCEPTION

The causal theory of perception has been held by philosophers, psychologists, and scientists in one form or another from the pre-Socratic predecessors of Plato through the development of modern science in the seventeenth century by Boyle and Newton, to recent philosophical theories such as those held by Price, Broad and Russell. The theory has been based upon philosophical speculation combined with observation, upon the data of physiology, and recently upon the results of modern physics. Thus, it shares a wide diversity of justification. In its precise formulation, the theory advocated by this tradition of thinkers is best stated by Price. Working within the philosophical framework of the sense-datum theory, Price investigates two main aspects of the problem of perception: the way in which knowledge is obtained and the relation of 'belonging to' holding between sensible qualities and physical objects. In these terms, the causal theory maintains (1) "that in the case of all sense-data (not merely visual and tactual) 'belonging to' simply means *being caused by*, so that 'M is present to my senses' will be equivalent to 'M causes a sense-datum with which I am acquainted'," and (2) "that perceptual consciousness is fundamentally an *inference* from effect to cause."[1] Attributing the theory to science, Russell describes it in almost identical terms.

Science holds that, when we 'see the sun' there is a process, starting from the sun, traversing the space between the sun and the eye, changing its character when it reaches the eye, changing its character again in the optic nerve and the brain, and finally producing the event which we call 'seeing the sun'. Our knowledge of the sun thus becomes inferential; our direct knowledge is of an event which is, in some sense, 'in us'.[2]

In *Knowledge, Its Scope and Limits*, he identifies the causal theory with the empirical, as opposed to the idealist, theory.

[1] Price, *Perception*, p. 66.
[2] *The Analysis of Matter*, p. 197.

"According to the empirical theory, some continuous chain of causation leads from the object to the percipient, and what is called 'perceiving' the object is the last link in this chain, or rather the last before the chain begins to lead out of the percipient's body instead of into it." (p. 195) Köhler, working within the context of physiological psychology, also accepts the causal theory. "An object sends out messages which stimulate sense organs. Thereupon other messages begin to travel through nerves towards the brain, and, if this brain is functioning normally, a percept emerges. This is a long chain of processes."[1] Later, Köhler stresses the inferential character of our perceptual knowledge. "Thus, our approach to the physical domain will under all circumstances consist of inferences which we draw from the observation of certain percepts, or, perhaps, also from other experiences; it will always be a procedure of *construction*."[2] The more plausible part of this theory is best expressed by Price's first statement, for it is clear, both from psychological investigation and ordinary reflexion upon the problem, that we do not gain our knowledge of what we normally take to be the external world through actual inferential operations. Köhler does not claim that perceptual knowledge of the ordinary world is inferential, but he insists that inference does characterize scientific knowledge. He sharply distinguishes between the phenomenal and the physical worlds. He readily agrees that our knowledge of the world of physics is inferred and constructed, but the 'things' we know first are not these abstract and shadowy objects but rather the immediate Gestalten of our direct experiences. These are not the result of conscious inferences: they emerge from the spontaneous activity of the nervous system.[3] Many philosophers and psychologists have charged the sense-datum theory with being post-analytic since they believe the evidence shows conclusively that actual awareness does not move from discrete sense-data to physical objects. The attempt to make perceptual consciousness an actual inference from effect to cause is even more clearly an analytical or epistemological dissection of what really takes place. Perceptual knowledge of the ordinary world may be

---

[1] *The Place of Value in a World of Fact*, pp. 109–110.
[2] *Ibid.*, p. 142. Cf. *Gestalt Psychology*, pp. 16–17.
[3] Cf. *Gestalt Psychology*, pp. 16–17; 117–126.

defended logically by resorting to this aspect of the causal theory. Even more plausibly, our knowledge of the world of physics may very well be an inference and the product of construction, as Köhler and Russell maintain. But no one can doubt that the causal theory is wrong if it is meant to ascribe any such constructive inferences to perceptual awareness.

> I think we must admit that, historically speaking, none of us reaches the belief in matter by inference, but that we all had it from the beginning: historically we all begin by *taking for granted* that visual and tactual sense-data are somehow constituents of the surfaces of material things.[1]

As a matter of fact, no one has ever seriously held the doctrine that perceptual consciousness is characterized by an inference. In every case, it seems to have been reflexions upon the difference between the world as first accepted and known through immediate experience and the world later constructed or supposed existing beyond and behind this immediate phenomenal world, which has led men to talk of inferences in respect to perceptual knowledge. The world of modern science has become much too abstract and strange to be immediately intuited. The very processes used by the physicists in arriving at their conclusions rest upon inferences from data to their causes. It is, then, most important to specify what world one has in mind when the inferential aspect of the causal theory is under discussion. Some sort of separation between phenomenal and physical worlds, between the worlds of sense-data and physical objects must be made before any acceptable meaning can be given to this aspect of the causal theory.[2]

It is precisely this dualism of phenomenal and physical worlds which the operationalists and the positivists have tried to eliminate from the interpretation of science on the grounds that it leads to a non-empirical analysis. Russell insists that the causal theory is an empirical theory but the curious result of this theory is a two-world ontology which sanctions a non-sensible

---

[1] Price, *op. cit.*, p. 99.

[2] One modification needs to be made in Price's statement of the first part of this theory, a modification which he himself suggests. Obviously, in any act of perception, there are a multiplicity of causal factors leading to the final act of awareness. Some of these are constant and hold for all acts of perception regardless of the objects, those which Price calls 'standing conditions', e.g., source of light, the eye, the optic nerve. Others vary with the experience and account for the differences in our perceptions: these Price calls the 'differential conditions'. It is with the differential conditions only that the causal theory seeks to identify the sensible qualities.

ingredient in the account of the world. This dualism is clearly evidenced in one of the earliest and most detailed statements of the causal theory, i.e., Plato's account in the *Timaeus* and *Theaetetus*. For Plato, the theory of perception is built within a world which is ceaselessly moving. Both the physical object and the sense organ are in motion. The interaction of the two causes sensation. It is sensation (or perception) which produces the perceptual (as opposed to the physical) object. The phenomenal world, the world of ordinary perception, is caused by the interaction of sense organ and physical objects; the qualities of the perceptual world differ fundamentally from those of the physical world. The object as known does not reveal any of the objective, non-sensible characteristics which that object as cause possesses in itself. The properties of the physical object for Plato are described in terms of the four primary bodies, the geometrical solids. While discussing the sense quality 'hot', Plato indicates the inferential nature of our knowledge of the physical world.

We are all aware that the sensation of fire is a piercing one; and we may infer the fineness of the edges, the sharpness of the angles, the smallness of the particles, and the swiftness of the movement, all of which properties make fire energetic and trenchant, cleaving and piercing whatever it encounters.[1]

While discussing the qualities 'hard' and 'soft', the dualism of phenomenal and physical again appears. "A thing is yielding when it has a small base; the figure composed of square faces, having a firm standing, is most stubborn; so too is anything that is specially resistant because it is contracted to the greatest density."[2] In general, the causal explanation of the genesis of sense qualities presented in the *Timaeus* attests to the fact that the physical causes of qualities have properties of their own which are never sensed, since in each case, the taste, the sound, and the tactile quality which is sensed is correlated with certain very definite sub-microscopic structures in the physical causes external to the observer. A feature not mentioned in Price's

---

[1] *Timaeus*, 61DE, Cornford's translation.
[2] *Ibid.*, 62C. For a detailed discussion of Plato's theory of perception and of the dualism implicit in it, see my discussion, "The Ontological Status of Sense-Data in Plato's Theory of Perception," *Review of Metaphysics*, vol. 3, 1950, pp. 21–59.

initial formulation thus appears in Plato's form of the causal theory, i.e., the ascription of non-sensible qualities to the differential condition.

The corpuscular version of the causal theory in the seventeenth century, in the hands of Gassendi, Hobbes, and Locke on the side of philosophy and of Boyle on the side of the new science, postulated tiny particles as the cause of perception, particles too small to be observed. But in the seventeenth-century form the theory does not make sufficiently clear just how obscure the qualities of the particles in themselves are. Locke advocated a substance as the basic cause of qualities and insisted that we have no knowledge of the real essence of this cause, while at the same time insisting that certain sensible qualities do belong in more than a causal sense to their causes. While trying to dismiss all physiological considerations from his account in order to keep the analysis of perception on the phenomenological level (asserting that the nominal essence is quite sufficient for our needs), Locke includes in his *Essay Concerning Human Understanding* long passages devoted to showing that we have a real knowledge of some of the qualities of the physical objects causing the qualities which constitute the nominal essence. That is, we do have a direct experience of the primary qualities although we do not know what precise collection of these qualities constitutes the real essence of objects. There are even passages which suggest that it might be possible with better microscopes to penetrate to the real essence of things. But in the *Essay* the discussion of substance and its allied concepts does not usually proceed as if Locke had these mechanical difficulties of science in mind. The dualism implicit in Plato's version of the theory has become ambiguous in Locke's account and we find ourselves oscillating between a phenomenological attitude clearly presaging the operational theory of present-day science, and an ontological dualism with its sharp separation of phenomenal and physical realms.

In Boyle's account of the causal, corpuscular theory of perception, we do not find the troublesome concept of substance, although he makes use of (indeed probably suggested to Locke) the same distinction employed by Locke between qualities belonging to matter actually and those which are, like Plato's sensible qualities, merely results of the operation of standing and differential conditions. Decrying the scholastic attempt to explain perception in terms of occult causes, Boyle declares that

All sorts of qualities ... may be produced mechanically. I mean by such corporeal agents, as do not appear, either to work otherwise, then by vertue of the Motion, Size, Figure, and Contrivance of their own parts, (which attributes I call the Mechanical affections of Matter, because to them men willingly Referre the various operations of Mechanical Engines) or to Produce the new Qualities exhibited by those bodies, and their changes ... by changes in the Texture, or motion, or some other Mechanical affection of the Body wrought upon.[1]

Qualities like texture, motion, and size Boyle calls the "simpler and more primitive affections of Matter", while colors, heat, smells, etc., are secondary qualities. Like Locke, he believed the relation between these two sets of qualities is not only causal but logical in the sense that the secondary can be deduced from the primitive qualities. Despite the interesting fact that Boyle understood and used the now familiar language of dispositions, a language common to most phenomenalists and inherent in the scientific concept of operationalism, he applied this concept only to the secondary, or as he usually referred to them, the sensible qualities.

I do not deny, but that Bodies may be said, in a very favourable sense, to have those Qualities we call sensible, though there were no animals in the World: for a Body in that case may differ from those Bodies, which now are quite devoid of Qualities, in its having such a disposition of its constituent Corpuscles, that in case it were dully apply'd to the Sensory of an animal it would produce such a sensible Quality, which a Body of another Texture would not ...[2]

Again, "if there were no Sensitive Beings, those Bodies that are now the objects of our senses, would be but dispositively, if I may so speak, endow'd with Colours, tasts, and the like."[3]

The primary-secondary distinction of the seventeenth-century variant of the causal theory is preserved in contemporary sense-datum theory. The point of departure for this theory is usually a criterion of physicality, a list of properties requisite for a physical (as opposed to a sensible) object. Broad lists five such characteristics: the object must have a 'reasonable' duration and

[1] *The Origine of Formes and Qualities*, 1666 ed., pp. xi-xii.
[2] *Ibid.*, p. 47.
[3] *Ibid.*, p. 49.

temporal unity, be multiply accessible to different observers and to different sense modalities, and have primary as well as secondary qualities.[1] The primary qualities are assumed to belong to some part of the physical object which is independent of observers, that part which Broad terms the 'scientific object', the object of inference which is located in the constructed physical space. Price's physical occupant plays an analogous role in his version of the theory. Lovejoy has a similar list of physical qualities in his *The Revolt Against Dualism*. And Russell, the philosopher who has most influenced Eddington, insists that the atomic structure of physics is an important non-sensible component of the physical world.[2]

The role of inference in the sense-datum theories of perception emerges from this basic distinction between primary and secondary qualities, between the sensible and the scientific worlds; for it is a process of inference by means of which the advocates of this theory seek to progress from sense-data to physical objects. Working from the circumscribed area of perceptual space and private sense fields, the advocates of the sense-datum theory seek to find some basis for making logical inferences from sense-data to their physical causes. Price and Broad suggest that if the hypothesis of an independent, external world has an *a priori* finite probability, the evidence of sensation increases this probability. Russell argues that unless the hypothesis of isomorphism or structural similarity between percepts and physical objects is assumed, these inferences cannot be justified. In each instance, the kind of inference demanded is an inference from established premises to conclusion: the rationalistic urge for a deductive system, characteristic of certain continental philosophers (e.g., Leibniz, Kant, Hegel) and of modern science, motivates the analysis, although it is recognized that at best the system will be only probable. Comparing the method here involved to that employed in geometry, Russell explains the general characteristics of the needed inference.

---

[1] *The Mind and Its Place in Nature*, pp. 146–47.

[2] The dichotomy assumed in the definition or concept of reality held by sense-datum theorists raises some important problems both in the linguistic formulation of their dualistic systems and in the epistemological character required for such dualism. These difficulties, especially as they relate to Russell's and Eddington's theories, are discussed in chapter VII.

What we really have to begin with ... is hypotheses containing variables. In geometry, this procedure has become familiar. Instead of 'axioms', supposed to be 'true', we have the hypothesis that a set of entities (otherwise undefined) has certain enumerated properties. We proceed to prove that such a set of entities has the properties which constitute the propositions of Euclidean geometry, or of whatever other geometry may be occupying our attention. Generally it will be possible to chose many different sets of initial hypotheses which will all yield the same body of propositions; the choice between these sets is logically irrelevant, and can be guided only by aesthetic considerations. There is, however, considerable utility in the discovery of a few simple hypotheses which will yield the whole of some deductive system, since it enables us to know what tests are necessary and sufficient in deciding whether some given set of entities satisfies the deductive system.[1]

Such a system is of course commonly established within the physical sciences as a means of relating the components,[2] but in the causal theory of perception, the same deductive structure has been sought from sense impressions to their causes. Russell has offered the most elaborate of such systems of deductive inferences. It is Russell's version of the attempted inference which reappears in Eddington's expression of the causal theory. Both Russell and Eddington embody in their versions the basic feature of the causal theory, what Price has called 'the Method of Correspondence'. This method proceeds on the assumption that there must be as much reality in the cause as in the effect. In the hands of Russell and Eddington, 'reality' in this assumption is interpreted as involving a close correspondence of cause (physical world) and effect (phenomenal world).

There must be a cause not merely for the existence of sense-data in general, but for all the particular detail of all the sense-data which we actually sense. It follows that *wherever we find differences in the sense data, there must be differences in the cause.*[3]

Neither Eddington nor Russell holds to the doctrine of resemblance between the private data of sense and the inferred entities assumed as their cause, but Eddington did wish to

---

[1] *Analysis of Matter*, p. 2.

[2] Meyerson, for example, describes scientific explanation in these terms: "L'explication consiste à montrer que, étant donné l'ensemble des antécédants, ce qui s'en est suivi pouvait en être inféré par déduction, n'en était que la conséquence logique." (p. 67, *De L'Explication dans les Sciences*). However, the strict operationalist would not accept Meyerson's definition of explanation as a search for causes. For example, Bridgman says that explanation consists "in reducing a situation to elements with which we are so familiar that we accept them, as a matter of course...." (*Logic of Modern Physics*, p. 37).

[3] Price, *op. cit.*, p. 74.

maintain that "An apple in the familiar story [the common sense view] has a counterpart in the external world; none of our familiar conceptions are appropriate to describe the nature of this counterpart, and we can only indicate it by a symbol such as X. But at any rate we can then say that the counterpart of two apples in the familiar story is two X's." (NPS, p. 23) The correspondence of features in the phenomenal with features in the physical world is what Russell and Eddington have called 'structural', a doctrine which I shall examine in more detail in the next chapter; but it should be clear from this brief reference to Eddington's doctrine of structure and its connection with his causal theory of perception that he is in the general philosophical tradition which accepts the causal theory together with some sort of ontological dualism. His operationalism was radically curtailed by his acceptance of this tradition.

  Besides advocating a doctrine of structure, Eddington also followed Russell in dividing knowledge into immediate and inferred; he insisted that all knowledge in the physical sciences must begin with what Russell has called 'momentary empiricism', with the individual experiences which we as experiencing subjects are aware of. These are the experiences which are both the most simple and the most certain.

  The only subject presented to me for study is the content of my consciousness. According to the usual description, this is a heterogeneous collection of sensations, emotions, conceptions, memories, etc. The raw materials of knowledge and the manufactured products of intellectual activity exist side by side in this collection. We wish to pick out the raw material – the primitive data unspoiled by the intervention of habitual forms of thought. (PPS, p. 195).

  Just as he emphasized the necessity for taking pointer-readings in context with other such readings, so he argued that,

  One sensation (divorced from knowledge already obtained by other sensations) tells us nothing; it does not even hint at anything outside the consciousness in which it occurs. The starting point of physical science is knowledge of *the group structure of a set of sensations* in a consciousness. (PPS, pp. 147–148)

  Those acquainted with Price's detailed study of perception will recognize here Price's family of sense-data from which we infer knowledge of the physical universe. One important feature, however, separates Eddington from his philosophical contemporaries at Oxford and Cambridge: where they have appealed in their attempted constructions to all sensible qualities, Eddington found that for science, there was too much overlapping among the senses, and he accordingly reduced them to one, that of vision. There is perhaps a curious anticipation of Eddington's reduction in the way in which prior analyses of perception have tended to concentrate upon visual data as in some sense more primary and more easy to work with, but the assumption usually was that what holds for vision holds also for the other sense modalities. Eddington's reduction diminished the complexity and intimacy of the relation between common sense and science; it also rendered the doctrine of structure more plausible. But another important accomplishment of this reduction was that it brought Eddington's perceptual theory into conformity with his teaching concerning pointer-readings as the data of science. The connection in question here is brought out clearly in *Space, Time and Gravitation*, where he discusses the analysis of space.

If we examine the nature of our observations, distinguishing what is actually seen from what is merely inferred, we find that, at least in all exact measurements, our knowledge is primarily built up of intersections of world-lines of two or more entities, that is to say their coincidences. For example, an electrician states that he has observed a current of 5 milliamperes. This is his inference: his actual observation was a *coincidence* of the image of a wire in his galvanometer with a division of a scale. (p. 87)

Seen coincidences of this sort constitute the data from which inferences about the physical world can be made.

Inference infects all knowledge which extends beyond the momentary contents of consciousness, whether it be on the macroscopic or microscopic level.

An electron is no more (and no less) hypothetical than a star. Nowadays we count electrons one by one in a Geiger counter, as we count the stars one by one on a photographic plate. In what sense can an electron be called more unobservable than a star? I am not sure whether I ought to say that I have seen an electron; but I have just the same doubt whether I have seen a star. If I have seen one, I have seen the other. I have seen a small disc of light surrounded by diffraction rings which has not the least resemblance to what a star is supposed to be ... Similarly in a Wilson expansion chamber I have seen a trail not in the least resembling what an electron is supposed to be; but the name 'electron' is given to the object in the physical world which has caused this trail to appear. (NPS, p.21)

But as regards the physical world, two radically distinct kinds of inferences are involved. If Eddington is to remain circumscribed within his operational attitude, 'physical reality' must be defined as the theme of physics and its various operations. The physical world is the world known by means of the tools of the physical scientists and is never to be thought of as distinct and separate from these tools. The controversy of whether electrons are real entities or only hypothetical factors guiding scientific research, should not arise within the operational attitude; for electrons on this view are just the results of certain operations involving observation in a Wilson cloud chamber. The status of these objects in themselves should not arise as a problem. Eddington's operationalism demands a phenomenalism in which 'reality' is reduced to the sum of the actual and possible visual sense-data derived from the momentary experiences of all actual and possible scientific observers. Unlike the dualist, the physical world for the operationalist is always immanent in experience. But since any excursion beyond the immediate awareness of observers is hypothetical, involving past experience and memory, inference plays its role even in the knowledge of this immanent world. It is, however, a relatively safe and mundane inference, one which can be verified by further experience. There are, however, many indications in the writings of Eddington which suggest that he found it necessary to make inferences of a bolder, more transcendent nature, from sense impressions to an independent, causal physical world. One of his favorite illustrations when discussing the causal theory is that of the brain as a great telephone exchange centre and the nerve fibres as wires extending from this centre to the edge of an external world.

When messages relating to a table are travelling in the nerves, the nerve-disturbance does not in the least resemble either the external table that originates the mental impression or the conception of the table that arises in consciousness. In the central clearing station the incoming messages are sorted and decoded... (NPW, p. 277)

Eddington here reminds us of Bergson: "The brain is nothing other than a kind of central telephone bureau, in my opinion: its role is to 'give communication', or to prepare for it."[1] But unlike Eddington who believed the central operator played an

[1] *Matter and Memory*, p. 26.

active part in determining the nature of the calls coming in,
Bergson conceived of the brain as an automatic exchange: "It
adds nothing to that which it receives." Selection does occur
for Bergson on the side of the observer, in terms of needs and
interests, but essentially Bergson's perceptual theory abolishes
the dualism of sense-datum and realist philosophies. Eddington
quite openly accepted this dualism. In *New Pathways in Science,*
using the simile of the common sense or familiar story teller and
the scientific narrator, he asserts that the data which the latter
has to work with are the impulses transmitted along the nerves:
"it is to this material that we must appeal if we wish to discover
the truth behind the story." (p. 4) The impressions supplied by
the nerve impulses (strictly, only those of the optic nerve)
constitute the base for inferences to and about the nature of the
causal factor behind these impressions.

> All that physical science can assert about the external world must be
> inferable from these. If there is any part of our conception of the physical
> universe which cannot have come to us in the form of nerve signals we
> must cut it out. As in a beleaguered city there spread circumstantial
> rumours of happenings in the world outside which cannot have been
> received from without, so in our minds there arise all sorts of conceptions
> of entities and phenomena in the external world which cannot have been
> transmitted to us from outside. (NPS, p. 4)

In short, Eddington argues that

> We are acquainted with an external world because its fibers run into
> our consciousness; it is only our own ends of the fibers that we actually
> know; from those ends we more or less successfully reconstruct the rest,
> as a palaeontologist reconstructs an extinct monster from its footprint.
> (NPW, p. 278)

The messages from the outside world are dressed up with
colour, extension, and permanence, factors which are not legitimate parts of the coded messages: as predicates of the external
world, they are assumptions made after the messages have arrived,
"for the transmitting mechanism is by its very nature incapable
of conveying such forms of conception." (NPS, p. 4) Even if
the objects of the external world do have such qualities, they
cannot be transmitted by nerve impulses. All knowledge of such
a world lying outside and beyond the nerve endings is indirect
and inferential, but some aspects of the story deciphered by the
mind can be seen to be false inferences simply by analyzing the

structure of the nerve fibers. By studying the physiology of the nervous system we can learn all we can possibly know without conjecture about the external world, but what we learn in this way is only the bare outline of the physical world. When an external object "raps on the door at the extremity of a nerve, you cannot put your head outside to see what is rapping. You cannot know more of its nature than that it must be such as to account for the delivery of the raps in their sequence." (*Ibid.*, p. 6) It is because of this inability to put our heads outside our nervous systems to inspect the stimuli, that knowledge of the stimulus is inferential. Perceptual knowledge of this world works from a cryptogram and certain suggested keys contributed by both the organism and the stimulus. Unless we wish to accept the solipsist position, we must believe in the existence of a stimulus. Indeed, good evidence can be offered to warrant this belief. But giving an account of the stimulus, even in outline only, is much more difficult; for various kinds of stimuli could account for our experiences. Eddington's doctrine of structure attempts to specify one type of stimulus as the external cause of our sensations.

Although it may be possible to interpret these assertions concerning the role of nervous impulses in physical knowledge in a phenomenalistic way, keeping the inferences entailed in the attempt to decode the messages restricted to the immanent operational world, it can be shown that Eddington did not succeed in divorcing himself from a dualist view of reality. In fact, Eddington's doctrine of structure involves this transcendent view. Moreover, even within an operational attitude, science cannot exclude all references to an independent, external world, especially when it starts, as Eddington does, with immediate experience and seeks to pass from this to the physical world. The operationalist who takes his position seriously and seeks to make it completely exhaustive would not concern himself with nerve impulses as carrying coded messages from an outside world. The nervous system is just one of the tools which such a scientist should include in his definition of the physical world; for, if the world is to be defined operationally, the nerves, just as much as the more complex acts of measuring, constitute the materials of the operations which, if the view is to be pursued thoroughly, must be

reflected in our definitions. All talk of the causal theory of perception would have to be abandoned: correlations between visual
impressions and operations, actual or possible, would replace this
theory. That Eddington did not make such a replacement
throughout his writings indicates either that he did not find it
wise to do so or else that he failed to shake off the last vestiges of
the older view. Dingle has suggested that many other scientists
are today still in this predicament of not knowing how to discard
all the older attitudes and beliefs. The phenomenalist-operational
language can be used to state the scientific story; but the significant question is how adequate and useful is the scientific story,
the scientific concept of reality? From the ordinary perceptual
point of view, it cannot be doubted that the causal theory is
correct in its general insistence upon a causal chain in perception
from object to observer, but it may be doubted whether acceptance of this aspect of the causal theory necessitates the further
acceptance of a sense-datum or physiological limitation of the
data provided by this causal chain. It may be that the physical
sciences cannot escape the constructionist view of reality built
out of discrete data, that a phenomenalist construction such as
Russell attempted in *The Analysis of Matter* is the minimum
ontology for science; but it is important to recognize that Eddington applies the language of construction to the world of
physics and not to the common-sense world. The distinction
between phenomenal and physical worlds which Köhler insists
upon is preserved by Eddington. Thus, from the start the scientific story departs from the story told by common sense. Distortion
of the familiar world is implicit in the procedure of the exact
sciences, for, in the first place, they exclude all data save the
visual data of pointer readings, and secondly, (in Eddington's
version at least) they seek to construct a world from discrete
physiological sensations conceived as carrying hidden meanings
of another world.

The approach to problems of perception and reality of the
causal theory has usually been oriented towards a two-world
theory, an appearance/reality distinction. The world of atoms
and electrons, in its Greek, seventeenth-century, or twentieth-
century varieties, has controlled and guided the subsequent
analysis of knowledge. Such an orientation is sharply differen-

tiated from that of certain other philosophers working in harmony with psychology and in revolt against the dualism of the causal theory. Bergson for example considered the causal theory to be involved in a basic mistake, of moving in the wrong direction: from the centre (the subject) to the periphery (the objects at the end of the nerve fibers). Eddington's switch-board operator collects the messages from the incoming nerves and attempts to read from them an account of the external world.

> Whence, then, comes that idea of an exterior world constructed ar-tificially, piece by piece with non-extended sensations about which we neither understand how they succeed in forming an extended surface nor how they project themselves outside our bodies?[1]

In order to account for the factor of externality and to bring his theory of perception into harmony with what he took to be psychological truth, Bergson found it more easy to initiate his analysis from the givenness of a multutide of presentations ('images'), out of which the observer's body becomes the focal point of perspective and in relation to which the individual's external world becomes oriented. Perception for Bergson is more action of the body upon and with its environment than it is a cognitive state. Material and immaterial, body and mind are not viewed as radically distinct and separate but, as a recent disciple of Whitehead has said,

> two modes of activity integrated in one actuality. The one, physical prehension, is a direct *rapport* with other actualities; the other, conceptual prehension, is a selective interpretative activity. Mind is thus not a separate entity 'decoding' messages received through the bodily nervous system, but an originative form of activity arising out of these physical activities, and through them in contact with the environing physical world.[2]

Working within the context of Husserl's phenomenological analysis, Merleau-Ponty has similarly stressed the need for a non-intellectualist approach to perception. In a detailed discus-sion of the various functions of the body in perception, he has urged (a) that the individual should be viewed as a unit express-ing at different times a corporeal and a mental activity; (b) that the body is not separate from the mind but rather a base from which action and knowledge extend; (c) that, arising in volitional

---

[1] *Matter and Memory*, p. 46.
[2] Emmet, D., *The Nature of Metaphysical Thinking*, p. 61.

and intentional acts, a world gradually emerges from our bodily motions, themselves impregnated with past experiences; and (d) that the individual's orientation in the world and conscious awareness of his objects is not an intellectual or a mechanistic process, but rather an unconscious organic relation of body-mind to multifarious objects comprising the physical world. Body movement leads to formation of habits, many of them unconscious, which provide unreasoned or unconscious points of orientation for the movements.[1] Opposed as much to the intellectualism of Descartes as to the sense-datum analysis of the empiricists, Merleau-Ponty insists that the sense-datum analysis violates the facts and distorts the notion of perception.

The classical notion of sensation was not a concept of reflexion, but a late product of thought directed towards objects, the final term of the representation of the world, the one furthest from the constitutive source, and for this reason the least clear. It is inevitable that in its general effort for objectification, science may end in presenting the human organism as a physical system in the presence of stimuli which are themselves defined by their physico-chemical properties, may seek to reconstruct perception on this basis and to close the cycle of scientific knowledge by discovering the laws by which knowledge itself is produced, by founding an objective science of subjectivity. But it is also inevitable that this attempt fails.[2]

The controls for this approach to perception are an initial faith in a monistic picture of man and his world and a reliance upon biology and psychology for supplying the data for analysis of knowledge. It is the stress upon the biological categories which has led me to term this approach 'organic phenomenalism,'[3] since the biological emphases lead to an organic analysis of knowing while the monistic faith is revealed in the phenomenalistic nature of the ontology which accompanies this particular approach.

The inescapable impression of what Whitehead has called the 'withness of the body' has led him and others to make the organic categories the key categories, and to try to develop a theory of sense perception in which the 'sensa' are derived by abstraction from physiological functionings defined in terms of 'feelings.'[4]

The causal theory of perception, in the hands of the sense-datum men in particular, builds the world from discrete, ir-

[1] *Phénoménologie de la Perception*, pp. 114–172. Cf. p. 104.
[2] *Ibid.*, pp. 17–18.
[3] See below pp. 127–129.
[4] Emmet, *op. cit.*, pp. 43–44.

reducible factors which exclude all references to 'feelings', 'adverbial modes', or the 'pre-scientific, pre-reflective level' of Husserl, Merleau-Ponty, and Whitehead, Both the starting point and the conclusion of these two 'stories' of perception and reality (that of organic phenomenalism and that of the causal theory) differ. The former is concerned to make its story as faithful an account of perception as can be supported by psychological research. The cognizing subject is taken in all his fullness. No reduction to Eddington's monocular creature is tolerated or to the more robust man of the sense-datum theory. The ontology supporting or supported by the organic phenomenalist approach is not allowed to be dualistic: it must be a phenomenalism of the total cognitive organism. The latter approach, that of Eddington and Russell, is committed to building an ontology with the abstract data of the exact sciences. Eddington insisted that the scientific story must not diverge entirely from that of common sense, but he stressed repeatedly the radical differences. With pointer readings as the ultimate data, the world of physics must be described in terms of coincidences of such readings. When the nervous system is introduced as a tool for extracting pointer readings from our experiences and when concern is shown for the world of stimuli outside the nervous system, the world of physics is committed to combining an operational description with transcendent descriptive propositions. Eddington strove to reduce the antithesis between his operationalism and his concern with the transcendent aspect of reality through his doctrine of structure; but it is clear that the antithesis is never fully reduced. The examination of his concept of reality which I make in a later chapter will show this antithesis at work on that level. We have already encountered the same tension in his claim that the nervous impulses carry coded messages, for within the operational-phenomenalist language it would be impossible to make such a claim. In going outside the boundaries of this restricted attitude, he has committed himself to many problems, the foremost of which is to explain on what grounds he can make the claim that the messages are coded in the nerves. He cannot make this claim categorically. At most this could be an hypothesis constructed either in terms of the end goal of science or else in relation with more general factors of human nature. Eddington recognized this difficulty and admitted that the general solution to the cryptogram is only probable; but what is the guide for making even this probable interpretation of the coded data of the nervous system? The guide is clearly his desire to keep the scientific story as closely as possible related to the story of common sense, consonant with the peculiar limitations of science. Behind all of Eddington's attempts to construct a philosophy of the physical sciences is the feeling that science must not stray too far from common sense, that the common sense picture of reality is in some way the ultimate measuring rod against which the pictures of science must be laid. Certain basic categories of thought which he believed were operative in common sense and science alike helped him to merge the two stories together. The deciphering of the messages carried along the nerves from the external world is guided by these categories of thought. But apparently the cryptogram is also decipherable in terms of the doctrine of structure, although the relation between this doctrine and the primitive forms of thought determining our concept of reality is symmetrical, the forms of thought leading us to find in our sensations evidence for the doctrine of structure and the doctrine lending support to the primitive concept of reality. Eddington's form of the causal theory, then, cannot be studied apart from discussions of his doctrine of structure, especially since the inference from impressions to external world rests upon this doctrine.

# CHAPTER III

# THE DOCTRINE OF STRUCTURE

The previous chapters have described Eddington's attempt to work out an interpretation of the physical sciences which would combine operationalism and its consequent subjectivism with the causal theory and its concomitant implication of a non-sensible, real world. I have argued that there is no contradiction in such an effort, that operationalism is a restricted doctrine whose validity and value are confined to the actual working of physicists, while the broader perspective which seeks to place the physical sciences in their context is appropriate to a full philosophical interpretation of those sciences. The physicist in his laboratory need not be concerned with the genesis of sensation or with its causal antecedents: his concern as a worker in the field of physics is only with the data of his machines and instruments. It is from these that he strives to construct a world. But very few physicists, of course, content themselves with such compartmentalized duties, with such narrow, inhuman reflections. It is in his concern with the origins of his data and their significance that the working scientist becomes a philosopher. We have seen how Köhler insists that even the working scientist must look beyond his data in order to impart meaning to them, the context within the phenomenal field constituting the base from which meaning arises for the pointer readings. Köhler had in mind those factors of our phenomenal field by which we tell a 'current' situation from a 'pressure' situation, but he also stresses the dynamic interpretative factor which the observer brings to his perception. In his concern with the causal theory and with the meaning of the physiological data of this theory, Eddington is in general agreement with Köhler's position, for both men are seeking some primitive foundation on which they can rest their theories of meaning. While Köhler is, as a psychologist, concerned with the way in which 'things' become selected and unified in the phenomenal field (more after the fashion of the organic phenomenalist),

Eddington as a philosophical physicist is more concerned with explaining the construction of 'things' in the scientific or, as Köhler would say, the trans-phenomenal field. But both men stress the role of the observer in these different kinds of unification procedures. Both are striving to place the source of the process of object discrimination in basic factors within the observer. It is in this endeavour to find a base for the segregation of things and the meanings attached to these perceptual processes which will be near the observer, in his own phenomenal field or in his mental structure, that Eddington reveals an interesting continuity between his operationalism and his acceptance of the causal theory of perception. Just as operationalism strives to restrict meaning to the manipulatory and verificatory procedures which can be encompassed by any observer, so the causal theory seeks to interpret the physical world in terms of the data available within the observer's own experience. Eddington's insistence that science starts with common sense and in some manner must remain faithful to common sense is a clear reflection of the same urge at work in his operationalist position: the urge to begin analysis with the familiar and near at hand. Both points of view – that of operationalism and that of the causal theory – take the observer as their points of departure. But whereas Dingle seeks to construct the whole of science upon the experiences of the subject (essentially, within a solipsistic experience),[1] thereby striving to prevent the philosophical interpretation of the sciences from violating his strict operationalist theory of meaning and spilling over into the trans-phenomenal world, Eddington saw no reason for such a solipsistic interpretation. He took as his data not only the sense-data of momentary empiricism but the beliefs and interpretations placed upon those data by the ordinary man.

---

[1] In his *Through Science to Philosophy*, Dingle seeks to analyze the whole of the physical sciences in terms of the private data of the observer. In his obituary on Eddington (*Proc. Phys. Soc.* vol. 57) Dingle argues that the truth towards which Eddington was working in the latter portions of *The Philosophy of Physical Science* was "that since the ultimate evidence for any assertion at all one makes about the world must lie in his own experience, ... a systematic deduction from evidence to conclusion must necessarily begin with the experience one has one's self." (p. 248) Although Dingle admits an interpretative factor contributed by reason, he considers the experiences which reason interprets in their narrowest sense, as atomic sense-datum experiences. Unlike Eddington, Dingle is satisfied with a coherence theory of truth.

In his Gifford lectures, Eddington stressed the contrast between
the common sense picture of reality and the scientific account.
The same contrast appears as the basic theme in his *New Pathways in Science,* and it has since become commonplace to point
out this distinction. The weight of our ordinary experience and
of our interpretations of them in our unphilosophical moments
has been so great that even science has not been able to deny
these interpretations without offering elaborate apologies. Almost
the whole of contemporary British philosophy pays similar homage to common sense or ordinary speech. But what sort of
homage can be given to our ordinary experiences and beliefs by
science when both the data (especially in the exact sciences) and
the final picture of reality differ so radically and seem to contain
so little that is familiar? The problem has become aggravated in
modern physics, since the cleavage of the two worlds has been
made more abrupt than before.

> Until recently there was a much closer linkage; the physicist used to
> borrow the raw material of his world from the familiar world, but he does
> so no longer. His raw materials are aether, electrons, quanta, potentials,
> Hamiltonian functions, etc., and he is nowadays scrupulously careful to
> guard these from contamination by conceptions borrowed from the other
> world. (NPW, p. xv)

Eddington admitted that ultimately the two worlds are just
"two aspects or two interpretations of one and the same world,"
that the scientific account does take its point of departure from
common sense (although it moves rapidly away) and "in the end
it must return to the familiar world." In the conclusion to his
*The Mathematical Theory of Relativity,* he argues that "we have
been engaged in *world-building* – the construction of a world
which shall operate under the same laws as the natural world
around us." (p. 237) He goes on to enunciate, almost as a rule for
the scientific process of world building, its faithfulness to the
common sense story.

> The world which we have to build from the crude material is the world
> of perception, and the process of building must depend on the nature of the
> percipient. Many things may be built out of $G_{\mu\nu}$, but they will only appear
> in the perceptual world if the percipient is interested in them. We cannot
> exclude the consideration of what kind of things are likely to appeal to the
> percipient. The building process of the mathematical theory must keep

step with that process by which the mind of the percipient endows with vivid qualities certain selected structural properties of the world. (pp. 237-38)

But Eddington believed that "the process by which the external world of physics is transformed into a world of familiar acquaintance in human consciousness is outside the scope of physics." (NPW, p. xiv) Even outside physics, in metaphysics or general philosophy, the translation could not be complete, for admittedly not every factor in the scientific account can find a counterpart in familiar experience. There are many untranslatable idioms. The constructed entities such as electrons, quanta, or potentials were formulated in response to definite problems confronted by physicists once they had already passed a long way from the common sense world. Eddington believed that there is some sort of counterpart in the familiar world of the larger group entities of science, such as the scientific table described in his Gifford lectures. Here a translation could be made, but in the many other instances no counterparts exist and hence no translation could be made.[1] In brief, it is what Eddington calls the 'simpler elements of the scientific world' (energy, stress, electrons, etc.) which lack counterparts and which constitute the irreducible remainders in any attempted translation. But even though every element in the scientific story does not have a counterpart in the familiar story (to try to force such a correspondence is, Eddington believed, to fall into the error of explaining science by means of mechanical models), the scientific as well as the familiar story is controlled by certain interpretative patterns of the mind. Since the end product in both cases is the "peculiar interpretation of the code messages transmitted along the nerves" into the mind, since the same basic categories of interpretation are at work in both stories, the relation between the two worlds would seem to

[1] Cf. Dingle's statement in his review of Russell's *Physics and Experience* (*Proc. Phys. Soc.*, vol. 59, p. 511). "It is undoubtedly right to say that the physical world is brought into being in order to account for the very dissimilar world of percepts, but it is much less accurate to say that a particular element of the physical world is brought into being in order to account for a particular percept... All problems of physics are restricted to the physical world alone, percepts acting merely as a sort of Clerk of Works to ensure that the world is built according to specification, and all relevant problems of epistemology would seem to be world from the whole assemblage of percepts i.e., with the whole relation of concerned with the principles of inferring or constructing any kind of physical specification to building, and not with the connection between each brick and its correlative cause."

arise by science starting with common sense, refining its process of analysis so far that it constructs entities far removed from the starting point, but entities which are taken as aspects (structural features) of an objective world. The process back to the familiar world is made by the realization that the interpretations science has placed upon its data have themselves been directed by specific categories at work as much in the familiar as in the scientific worlds.

However much Eddington insists that the translation from scientific to familiar worlds lies outside the scope of physics, if his doctrine of interpretative categories is correct, the germ of the required translation is already contained in the scientific account. We have only to find the dictionary which will translate the terminology and concepts embodying these basic interpretative categories from one language to the other. Unity and permanence, for example, will inevitably be a feature of both accounts, although the language and concepts used to express these features will differ. Such a task of translation would indeed be beyond the scope of physics, but Eddington insisted that science aims at constructing a world which shall be symbolic of the world of commonplace experience.

> In the world of physics we watch a shadowgraph performance of the drama of familiar life. The shadow of my elbow rests on the shadow table as the shadow ink flows over the shadow paper. It is all symbolic, and as a symbol the physicist leaves it. (NPW, p. xvi)

If this is in fact one of the aims of science – to be symbolic of the familiar world – then the very aim of science consists in part of laying the foundation for the translation. Eddington's theory of symbols has, in fact, a double allegiance: scientific knowledge is symbolic both of the familiar world of common sense and of the objective world. Eddington formulated the doctrine of structure as a way of showing the objectivity of science, of showing how the scientific story does reveal some information concerning the external world. But the external world of common sense is not, on Eddington's view, very remote from the external world of science. Thus it is through the doctrine of structure that Eddington strove to achieve two goals: chiefly, to establish the objectivity of scientific knowledge, but also to establish the union between common sense and science.

This is, however, the end of the story and we have not as yet examined and made precise the doctrine of structure itself. An initial difficulty confronts the doctrine which warrants some discussion. The relation between physics and ordinary experience for Eddington is two-fold. Physics must be faithful to some extent to the beliefs and interpretations of experience inherent in common sense, but physics is also dependent upon ordinary perceptual experience to collect and note the data which it uses. Even though Eddington reduced the relevant data to visual pointer readings, and even though he recognized that physics has very little to do with ordinary perceptual experiences, he pointed out that it is the familiar story teller who reads the meters of the experiments. Thus, if we are to begin at the beginning we must inquire why we trust the story teller's information about galvanometers in spite of his general untrustworthiness. (NPS, p. 3) Eddington answered this question as follows:

I think the answer is that the truth of the story is not the point in question; the physicist is concerned only with the scraps of cipher contained in it. The galvanometer is a device for leading the story into situations in which the underlying cipher becomes less baffling to interpret; it is not a bridle on the story teller's imagination. (NPS, p. 11)

Eddington is not raising the ordinary doubts concerning the veracity of sense perception. He is not asking why it is that we believe the familiar story teller can read a meter correctly but report incorrectly concerning his more common perceptions. His question has a double reference: he is asking why pointer reading percepts are thought to be relevant to the scientist's aim as well as how percepts of this sort can be evidence for inferences to the external world. His answer to both aspects of the question is that pointer reading percepts allow some basis for constructing a structural knowledge of the external world. But this answer is valid only after we have accepted and made intelligible the doctrine of structure. Moreover the question raised by Eddington has a more general formulation.

Every empiricist holds that our knowledge as to matters of fact is derived from perception, but if physics is true there must be so little resemblance between our percepts and their external causes that it is difficult to see how, from percepts, we can acquire a knowledge of external objects.[1]

[1] Russell, *Knowledge, Its Scope and Limits*, p. 197.

The problem is further complicated by the fact that historically science started from the naive realist position of common sense, from the belief that external objects are exactly as they seem to ordinary experience, but science concludes by rejecting this familiar belief. In terms of science and the causal theory of perception, the observer does not experience directly the objects but only the effects of the object on him.

> Thus, science seems to be at war with itself: when it most means to be objective, it finds itself plunged into subjectivity against its will. Naïve realism leads to physics, and physics, if true, shows that naïve realism is false. Therefore, naïve realism, if true, is false; therefore, it is false.[1]

Again the problem is two-fold: (1) how can percepts, even of Eddington's restricted variety, serve as bases for inferences to external objects, since they are causally removed from their supposed objects and are not resemblances of them; and (2) how can such percepts be accepted as valid when science ends by rejecting what they initially report.

Even when we accept the causal theory and its consequent assertion of the existence of an object at the far end of the causal chain, we have still to state how it is that we can know the theory to be true or know that unsensed physical objects do exist as the originators of the percepts we receive. A simple empiricism (what Russell has termed 'pure empiricism') such as Dingle exposes cannot, of course, solve this problem. In fact, the problem would never arise; Dingle argues that we are not entitled to any such assumption of an unsensed external world as Russell and Eddington make. By abandoning the language of construction for that of inference he thinks Russell creates the problem he sets out to solve.[2] Dingle will have nothing to do with any position which involves him in a non-operationalist theory of meaning. One of the major weaknesses of Eddington's philosophical interpretation of the physical sciences is not, as Dingle would have it, Eddington's departure from a rigid operationalist attitude, but his failure to develop the details of the non-operationalist theory of meaning entailed in his dualistic position. Problems of meaning are crucial within the working sphere of the practising scientist,

---

[1] Russell, *Inquiry into Meaning and Truth*, p. 15.
[2] In his review of Russell's *Physics and Experience*, *Proc. Phys. Soc.*, vol. 59, pp. 509–11.

but they are of even greater importance when we attempt to place the sciences in their wider context. Some, such as Dingle and Bridgman, have argued that only an operationalist theory of meaning is legitimate within the sciences, but Eddington and Russell have not seen the necessity of so restricting their interpretations. But neither Russell nor Eddington has succeeded in presenting a theory of meaning to cover his dualistic ontology. Russell does recognize that as we pass from direct sense-datum statements of the sort, 'I am hot,' to memory statements of past experiences, and thence to statements like 'You are hot' or 'the sun is hot,' we pass further and further away from the possibility of finding verifiers for our statements. Where we allow statements (or beliefs) which cannot be verified, we have passed beyond the rules for an operational theory of meaning, since there are no operations one can perform to verify the assertion or in terms of which its meaning can be defined. In the case of those statements concerning unsensed causes of percepts, Eddington's hypothetico-observational criterion of meaning is of no use, for we are here beyond even the possibility of an observation which would define or verify the statement.

> Formally, whenever an assertion goes beyond my experience, the situation is this: inference leads me to 'there is an $x$ such that $\varphi x$', and this, if true, is true in virtue of an occurrence which would be asserted by '$\varphi a$.' But I know of no such occurrence. When I say 'I am hot', I am aware of the verifier, which is my hotness. When I say 'you are hot' or 'the sun is hot', I am not aware of the verifier.[1]

Russell believes the only way we can define the truth, and, it would follow, the meaning of statements going beyond even the possibility of verification, is in terms of basic existence statements which are in some sense derived from experience. But, since in no sense do we experience the cause of percepts, it is clear that something more is needed to establish the meaning or truth of such assertions and beliefs. The inference from sense-data to their causes cannot be made solely on the grounds of basic propositions or beliefs defined in terms of direct experience. Eddington did not realize sufficiently the need for explicating the meaning of his non-operationalist assertions. He and Russell both overlooked the importance of the problem of meaning in their dualistic

[1] Russell, *Meaning and Truth*, p. 232.

epistemology. The one place where Russell deals extensively with problems of meaning, in the *Inquiry into Meaning and Truth*, purports to rest all meaning upon basic propositions. It can be shown, however, that trans-phenomenal statements (statements whose meaning refers beyond the given, to some hypothetical but unattainable verifiers) cannot be made to fit into this meaning scheme.[1] One area of the relation between common sense and physics is thus left unanalyzed by both Russell and Eddington. In committing themselves to the task of squaring science with the beliefs and interpretations of common sense concerning an external world, they have committed themselves to explaining why science finds it necessary to preserve this belief in an external world by denying the validity of naive realism. It is in this process of showing naive realism to be false that science, on Russell's and Eddington's interpretation, introduces a level of meaning not present in its most radical form in the common sense picture.[2]

While Eddington and Russell failed to work out the details of the genesis of such a level of meaning or its justification they are aware of the necessity of providing some grounds to validate the type of meaning which they introduce. The doctrine of structure formulated by both men is meant to explain the inference from percepts to causes as well as Russell's query about the validity of a science which begins with naive realism and ends by concluding that naive realism is false. Explaining the inferences entailed in this view also helps to account for the meaning these inferences assume. If true, the doctrine of structure would validate the non-operationalist theory of meaning of these statements and inferences, since then basic propositions and sense-data would be the grounds in terms of which the meaning and truth of these existential statements and inferences of the causal theory would be defined. The meaning of any statement involving the assertion of unsensed causes required by the causal theory would be defined in terms of a series of sense-datum or percept

---

[1] I discuss this problem at length in chapter VII.

[2] The status of memory propositions and existential propositions asserting the existence of non-experienceable entities in the external world, the former of which is present in common sense, the latter in science, is logically the same, but the latter is of a more crucial nature. Memory propositions, when correct, are of events or experiences once enjoyed or endured, but statements about unsensed physical causes, even if true, are not about experienced or even experienceable events or objects.

statements. That these statements would define the meaning of the transcendent assertions would be at least partially stated in virtue of the doctrine of structure, which posits a structural relation between cause and percept. If percepts are to be the source of our knowledge of the external world, and if an external world distinct and different from our percepts does in fact exist, then there must be some justification for making inferences from effects (percepts) to their causes (physical objects). The major basis for justifying this inference and for providing a meaning criterion for the statements expressing these inferences which Russell and Eddington have suggested is the presupposition of Price's method of correspondence. Russell describes this presupposition as that of 'the separable causal chains'.

> I can see at the present moment various things – sheets of paper, books, trees, walls, and clouds. If the separateness of these things in my visual field is to correspond to a physical separateness, each of them must start its own causal chain, arriving at my eye without much interference from the others.[1]

Of course, physical separateness is precisely what remains to be proved if the supposition of structure is to be supported. In his *Analysis of Matter*, Russell expresses the doctrine in more precise terms.

> We assume that differences in percepts imply differences in stimuli-i.e., if a person hears two sounds at once, or sees two colours at once, two physically different stimuli have reached his ear or his eye. This principle, together with spatio-temporal continuity, suffices to give a great deal of knowledge as to the *structure* of stimuli. Their intrinsic characters, it is true, must remain unknown, ... What we assume is formally, something like this: there is a roughly one-one relation between stimulus and percept – i.e. ,between the events just outside the sense-organ the and event which we call a perception. This enables us to infer certain mathematical properties of the stimulus when we know the percept, and conversely. ... (pp. 226–227).

For every difference in our perceptual field there is a corresponding difference in the physical field causing the percepts, and vice versa; where there is identity of percepts, as between those of several individuals, the presupposition invoked by Russell and Eddington allows us to infer an identity of causal factors. "If two bodies are of identical structure as regards the complex of interval relations, they will be exactly similar as regards observational

---

[1] *Knowledge, Its Scope and Limits*, p. 206.

properties, if our fundamental hypothesis is true." (MTR, p. 41) However, since we are excluded from an intimate knowledge of the physical causes of sensation, we cannot state in very definite terms just what is identical or different in the physical field. Likewise, in order to make sense of the presupposition of separable causal chains, some content must be given to the physical field, however indefinite. Thus, the knowledge we have of the external world is, in Eddington's language, shadowy and symbolic. 'Structure' becomes the key phrase for explaining the kind of knowledge we do have of this unsensed causal realm.

> The investigation of the external world is a quest for *structure* rather than substance. A structure can best be represented as a complex of relations and relata; and in conformity with this we endeavour to reduce the phenomena to their expressions in terms of the relations which we call intervals and the relata which we call events. (MTR, p. 41)

Understanding of the external world, while never complete because of the skeletal nature of such structural or formal knowledge, is still sufficient to yield some insights. "To understand the phenomena of the physical world it is necessary to know the equations which the symbols obey but not the nature of that which is being symbolised."[1] For Eddington as well as for Russell, the doctrine of structure seems to be the answer to two problems: what to do with the formalized, mathematical knowledge of the exact sciences, in order to relate it to the physical world it is supposed to describe, and how to justify the beliefs of common sense in an independent external world. The doctrine accomplishes both tasks of explanation, for it makes sense of the existential inference from sense-data to physical objects as well as lends significance to the mathematical formalizations of the exact sciences. The language of construction common among those who philosophize about the exact sciences is one way of attempting to build a world out of the mathematical data of measurements. With such materials at our disposal for the task of construction, it is no surprise that our result is a formal pattern or structure bearing little resemblance to the world of common sense. But through the doctrine of structure, Eddington seeks to relate this formalized world constructed by the exact sciences to the objective world believed in by common sense.

[1] Eddington, *Science and the Unseen World*, p. 20.

The same series of operations will naturally manufacture the same result when world-conditions are the same, and different results when they are different. ... The study of physical quantities, although they are the results of our own operations (actual or potential), gives us some kind of knowledge of the world conditions, since the same operations will give different results in different world conditions. It seems that this indirect knowledge is all that we can ever attain, and that it is only through its influences on such operations that we can represent to ourselves a 'condition of the world.' (MTR, pp. 2–3)

Quantitative measurements are controlled in part by the world conditions which we never know directly, while our final interpretation of these hidden conditions is itself directed by our interpretative categories.

In his *Les Mathématiques et la Réalité*, Professor F. Gonseth has advanced very similar views. He begins by suggesting that "our ideas of the world carry the mark of our own mental structure, in the same way that the personality, the artistic structure of a painter, is rediscovered in his style." (p. 53) He stresses the schematical, or as Eddington says, the structural character of our objective knowledge. He compares it to the knowledge an explorer has of an unknown region in terms of location points which he lays down, permitting him to traverse the area. "In other words, reality *as we perceive it*, is a construction, more or less autonomous, of our mind, the essential ends of which are to make action possible." (p. 54) Like Husserl (in *Erfahrung und Urteil*) in trying to define the meaning of 'object,' Gonseth strives to recover the primitive, pre-judgmental experiences, but he insists upon the approximate and schematic character of our knowledge. (p. 161) The world conditions of Eddington's *Mathematical Theory of Relativity* are the features of the objective world which our knowledge strives to schematize. Eddington insisted that our interpretations of the world conditions behind our exact measurements are only probable; it is largely conjecture which leads to any one plan of schematism. Awareness of the uncertainty of such interpretations led him on occasions into a kind of conventionalist attitude reminiscent of Poincaré.

The physical quantity is the measure-number of a world-condition in some code; we can not assert that a code is right or wrong, or that a measure-number is real or unreal; what we require is that the code should be the accepted code, and the measure-number the number in current use. (MTR, p. 4)

At other times he admitted that it was possible to derive more than one schema or structure from our data, but however problematic our deciphering may be, he was convinced that it contains some degree of truth, that it reflects to some extent the real structure of the world.

In a large measure, this conclusion of Eddington's concerning the reflexion of the objective world structure in our scientific data is a matter of what Santayana would call animal faith, the same faith which motivates common sense in its belief in an external world. Eddington would not be satisfied with anything short of a science which could be related to this fundamental belief in an objective, external world. However much he was drawn towards the operationalist point of view, however much he desired to content himself with the language of construction, it was the language of inference which emerged, once the task of construction had been carried out. There is no definitive proof offered by Eddington to substantiate his belief in the structural representation of our perceptions of an external world; but, aware of the need of justifying this doctrine, he did offer some arguments in support of his belief. There are two sides to the problem: the general meaning of the term 'structure' and the reflexion of structure in different systems of objects, events, or experiences. The problem of justifying the doctrine belongs to the latter side. Both Russell and Eddington approach this problem by calling attention to the similarity of structure in the perceptual fields of several individuals, to the fact that many individuals can have percepts related in the same way as to position, order, and size. Pushing aside the question of comparing my sensations with some one else's sensations, Eddington argues,

It is possible for a group of sensations in my mind to have the same structure as a group of sensations in your mind. It is possible also that a group of entities which are not sensations in anyone's mind, associated together by relations of which we can form no conception, may have this same structure. We can therefore have structural knowledge of that which is outside everyone's mind. (PPS, p. 142)

The possibility of a structural knowledge of the external world is thus established. If this possibility could be realized, the doctrine of structure could overcome the epistemological problem of external knowledge which a dualist ontology must face; for, "so long as the knowledge is confined to assertions of structure, it is not tied down to any particular realm of content." (PPS, p. 143) Eddington uses this aspect of the doctrine of structure as support for a theory of neutral monism, where mind and matter are no longer in sharp contrast.

> The recognition that physical knowledge is structural knowledge abolishes all dualism of consciousness and matter. Dualism depends on the belief that we find in the external world something of a nature incommensurable with what we find in consciousness; but all that physical science reveals to us in the external world is group-structure, and group-structure is also to be found in consciousness. (PPS, p. 150)[1]

'Structure' functions similarly to the scholastic concept of 'form' by means of which Aristotle and St. Thomas claimed we have a knowledge of the external world. The form, phantasm, or species for St. Thomas is just the essentials of the physical object existing in the mind, devoid of matter. The sensible form, like Eddington's structure, is both in the object and in the observer's experience.

> In the same way, the sensible form is in one way in the thing which is external to the soul, and in another way in the senses, which receive the forms of sensible things without receiving matter, such as color of gold without receiving gold. So, too, the intellect, according to its own mode, receives under conditions of immateriality and immobility the species of material and movable bodies; for the received is in the receiver according to the mode of the receiver.[2]

There is no problem of how a material phantasm can enter an immaterial mind, since the phantasm is interpreted formally and is capable of existing both in matter and in mind. In the very same fashion, Eddington and Russell explain the problem of knowledge in terms of structure, the structure now being taken

---

[1] Cf. *Nature of the Physical World*, p. 280: "Consciousness is not sharply defined, but fades into subconsciousness; and beyond that we must postulate something indefinite but yet continuous with our mental nature. This, I take to be the worldstuff. We liken it to our conscious feelings because, now that we are convinced of the formal and symbolic character of the entities of physics, there is nothing else to liken it to". Also, in *New Pathways in Science*, he says that, taking the external world as the solution of the cryptogram of our sensory experiences, this solution is not associated with any one form of existence. (p. 18). Eddington's neutral monism is of course very similiar to that developed by Russell in the *Analysis of Mind*.

[2] *Summa Theologica*, Q. 84, Art. 1.

as the formal factor capable of exemplification in many diverse events and objects. The information imparted by a sound, for example, is composed of (a) "a general concept of a heard-noise, i.e., a concept of something of similar nature to my own awareness of noises," and (b) "a structural concept of a heard-noise, i. e., a part of the structure of the physical universe which we describe as an electrically disturbed terminal of an auditory nerve." (PPS, p. 149) Of these two concepts, Eddington significantly remarks that "the one refers to what it is in itself, the other refers to what it is as a constituent of the structure known as the physical universe." (*Ibid.*, p. 150) The same event belongs to two realms of existence and it is identical in both realms in all essential features, i.e., in its formal properties. We become acquainted with this ability of structure to exist as one type or another by comparing our sensible experiences with other experiences of different individuals. The first step towards structural knowledge is "a comparison of sensations in one consciousness," and this is in turn transferred to comparison between several minds. (PPS, p. 198) Like Russell, Eddington offers the fact of similar structures in the experiences of many individuals as evidence of an external world, both men invoking the fundamental presupposition of the method of correspondence as a means of characterizing this world.

Communicability of knowledge is also made to rest upon the similarity of structure in the experiences of different persons. Out of the comparisons of our own sensations we come to "extract a pattern of interlocking, which can be described mathematically and represents structural knowledge of the sensory content of the consciousness studied." (PPS, p. 207) When we discuss this pattern of interlocking with other individuals, we discover similar patterns in their experiences.

The problem then arises, How are we to represent this interdependence? We may begin with the simple case in which the same structure is found in nearly all consciousnesses with which we can communicate, for example, the structure of the visual sensation which arises when we look at a constellation in the starry heavens. We reject the idea that the occurrence of this highly specialized structure in so many consciousnesses is a coincidence, and thereby commit ourselves to the hypothesis that the many similar structures are reproductions of one original structure. This is the germ of the idea of causation. (PPS, p. 208)

The similar structure is then attributed to an external world. The external world is introduced in order to account for the similar patterns of interlocking sensations in different individuals' experiences. "Since the external world is introduced as a receptacle of structure, our knowledge of it is limited to structural knowledge"; the possibility of a Berkelian solution, where the common patterns of experience are attributed to the causative power of God, or Leibniz's monadistic solution, is ruled out as pure coincidence.

We do not, to begin with, put forward any theory as to *how* the original structure in the external world comes to be reproduced as a structure of sensations in consciousness; we merely recognize that, ruling out coincidence, the occurrence of the same structure in many consciousnesses is a sign that an original structure exists in a realm outside those consciousnesses. (PPS, p. 209)

Russell calls attention to the implicit presence of the same sort of argument in other situations, as when we infer a causal connection between our shadow and our body after having noted identity of pattern in the movements of shadow and body. The same argument is present in the 'brides-in-the-bath' murder cases.

A number of middle-aged ladies in different parts of the country, after marrying and insuring their lives in favor of their husbands, mysteriously died in their baths. The identity of structure between these different events led to the assumption of a common causal origin; this origin was found to be Mr. Smith, who was duly hanged.[1]

Russell fails to call attention to the possibility that the common causal origin of these deaths need not all have been the same individual, but he is correct in arguing that under normal conditions we do look for a common motive underlying a series of events of this nature. The inference from common pattern of events to common cause or common motive is not a necessary one; our conviction stems from past experience where we have come to expect, as Hume remarked, the same antecedent for the same sort of event. Admittedly we do not have previous conjunctions of cause and effect between percepts and their supposed causes on which we can form expectations, the lack of which leaves open the possibility of some other explanation to

[1] *Knowledge, Its Scope and Limits*, p. 464.

account for observed phenomena; but these explanations (e.g., that of Berkeley or Leibniz) are further removed from the common sense belief which motivates those who advance the doctrine of structure. Russell believes that the many instances of common patterns in our experience having common causal ancestors lends some probability to the hypothesis when applied to our percepts and their supposed external causes. It is more difficult, though, to argue for similarly structured causes and effects than it is to argue for a common causal ancestor behind similar events. The doctrine of separable causal chains in the physical world is the presupposition of the doctrine of structure and as such cannot be proven. The arguments advanced in its favour by Russell and Eddington do not so much lend support for the doctrine as give meaning to it. They attempt to extend a concept which they believe has applications in our normal experience to an area of unexperienceable reality. For the final evaluation of the doctrine, we have to examine the illustrations offered in order to discover just what meaning they give to 'structure' when applied to this realm.

We have already seen that 'structure' is taken as the formal or logical property of objects or of a series of events. 'Order' is clearly one of the qualities subsumed under the head of 'structure.' Russell draws a number of illustrations from various fields all of which show the importance of order in the concept of structure. Beginning with logic, his main interest, Russell remarks that the logician is concerned with the formal relations between sentences or propositions. Sentences of the sort, 'Brutus killed Caesar,' or 'Socrates is Greek' are of the same logical structure in being dyadic-relation propositions. Other propositions are classed as hypothetical, categorical, etc., while logical relations of other types are catalogued as symmetrical, asymmetrical, etc. All of these are instances of what Russell understands by 'structure.' Another familiar illustration used by Russell and Wittgenstein is that of the map. If the area to be mapped is small, "so that the curvature of the earth can be neglected," the principle of mapping is simple:

East and west are represented by right and left, north and south by up and down, and all distances are reduced in the same proportion. It follows that from every statement about the map you can infer one about the district, and vice versa. ... These inferences are possible owing to identity of structure between the map and the district.[1]

Again order is important as defining the concept of 'structure,' both the order within each object as well as the order which can be drawn between the objects. For each significant component of the map there is a corresponding component in the area which is being mapped. The doctrine of separable causal chains is thus suggested, although there is no causal connection between the map and the area. The identity of pattern between map and area is a constructed identity, purposely planned so that the map can be of some use. But if the doctrine of structure is to operate in the physical world between object and percept, a similar sort of identity of pattern must exist. Russell works with those familiar instances where the identity of pattern between two different sorts of objects is clearly in evidence. The map illustration is reinforced by that of the gramophone.

It is obvious that it could not produce this music unless there were a certain identity of structure between it and the music, which can be exhibited by translating sound relations into space relations, or vice versa; e.g., what is nearer to the center on the record corresponds to what is later in time in the music. It is only because of the identity of structure that the record is able to cause the music.[2]

Translation from one medium to another could not take place unless there is a one-one relationship between the elements being translated and their translation. The order of relationships within the translated object must be duplicated in the translating medium.

A wireless set transforms electromagnetic waves into sound waves; a human organism transforms sound waves into auditory sensations. The electromagnetic waves and the sound waves have a certain similarity of structure, and so (we may assume) have the sound waves and the auditory sensations.[3]

In the case of sensations, the transfer of structure as order takes place on three levels: the actual sensation felt and experienced directly, the physiological correlates of the sensation, and the distant causal object at the far end of the chain. Neither Russell nor Eddington succeeds in making clear what other

[1] *Knowledge*, p. 253.
[2] *Ibid.*
[3] *Ibid.*, p. 254.

factors besides order comprise the identity of structure on these three levels. Russell repeatedly describes structural properties as logical relations, insisting that the notion of structure "is not applicable to classes, but only to relations or systems of relations."[1] He would seem to limit 'structure' in its application to perception, to the relations between sense-data and the relations between physical objects and their parts. His assumption of separable causal chains is meant to place on a probable basis the similarity of relations between sense-data comprising phenomenal objects and their physical object causes. But Russell does not fill in the variables, we do not know what components of sense-datum relations he means to correlate with their physical causes. He would not say, I think, that every discernible feature or component of our sense field has an isomorphic relation with features in the physical field. We are not told what is the allowable basis for segregating components in our perceptual field. A percept of a table can be broken down into various components, such as those of top and bottom, sides, colors, hardness; but I might also divide the perception into different segments. For instance, I might divide the top surface of the table percept into several areas. As Cassirer has remarked, the unity of objects can never be determined with exactitude on the sensuous level or without specific criteria in terms of which the beginning and the scope of objects is constructed.[2]

Such criteria will be found later to be at work in the systems of Russell and Eddington as unquestioned axioms or categories determining the very nature of their ontologies and hence of their epistomologies. In Russell's case, the failure to recognize the fundamental role played by these physical object criteria was due to two main causes: his concern with formulating the phenomenal levels of science and knowledge, and his emphasis upon the language of construction. The complexities of the task of construction lead easily to the belief that no extraneous concepts have moulded the final product. But even here, the vague beliefs of common sense are recognized as controlling factors in the construction. For example, the perceived space relations in private space are not identified

---

[1] *Analysis of Matter*, p. 249. Cf. *Principia Mathematica*, vol. II, part IV*, 150.
[2] *Substance and Function*, p. 120.

with those which physics assumes among the corresponding physical objects but they have a certain kind of correspondence with those relations. If we represent the position, for physics, of visible objects by polar co-ordinates, taking the percipient as origin, the two angular coordinates correspond to perceived relations among visual percepts, while the radius vector ... is inferred by means of causal laws. ... My point is that the relations which physics assumes in assigning angular co-ordinates are not identical with those which we perceive in the visual field, but merely correspond with them in a manner which preserves their logical (mathematical) properties.[1]

The selection of corresponding components is controlled, in other words, by the generally accepted concept of physical objects. The attempted construction has as its goal the representation of at least the outlines of the familiar world. But aside from order and certain mathematical properties, we are never told what are the logical properties of sense-data and physical objects which are related. The filling in, in more detail, of what is meant by logical or formal properties of objects, especially of percepts and their physical causes, might shed some light on this matter. As it stands, the concept of structure as applied to this area is left largely in the realm of analogy. We are told that the structural relation between percepts and physical objects is like that obtaining between the gramophone and the music or the map and the area mapped. By the very nature of the situation we cannot gain a direct knowledge of the two terms of the relation in the case of percepts and causes and hence cannot verify the hypothesis of structural similarity. This situation makes it difficult to render explicit the concept of structure as here applied.

Because of the indirectness of much of our knowledge, Eddington liked to speak of our structural knowledge as shadowy and symbolic. In fact, he divided the whole of knowledge into two great classes: intimate or direct and symbolic or indirect. But again, we never find in Eddington's writing any clear statement of what he understands by a 'symbol.' Emmet has pointed out the important distinction between symbols which name and those which describe, calling attention to the fact that in the latter case, as for instance in mercator projections or diagrams of the benzene ring, we have "a certain systematic correspondence

---

[1] *Analysis of Matter*, p. 252.

to the structure of that which they symbolize."[1] With symbols which name, as in the case of many non-pictorial symbols, this simple correspondence does not obtain. Descriptive signs are replaced by conventional signs without any obvious isomorphic correspondence. Does Eddington employ the term 'symbol' as descriptive or conventional sign? He seems to oscillate between the two or to endeavor to incorporate both types. A symbol for him must provide enough information in lieu of direct knowledge of that which is symbolized to allow us to perform operations as if we did have the immediate type of knowledge. The symbol of the physical world, whether it be the sense-datum or the mathematical equation, replaces the event or object symbolized. Stebbing has raised the obvious question of how Eddington could determine that any section of our physical knowledge is symbolic of something not known.

> How does the physicist ever succeed in making the discovery that physical laws are symbolic of 'true natures' that are inscrutable? How, finally, could the study of physics lead to the discovery that there are inscrutable natures which are 'some kind of counterpart in the external world, of that which is in our minds'?[2]

She points out the weakness of the basic analogy invoked by Russell and Eddington in their doctrines of structure.

> We know that a broadcast station has a nature of its own, that the announcer has a nature of his own, and that we shall not discover what either is simply by 'listening in.' But what reason is there for supposing that a chair has a nature utterly unlike how it behaves and what it is perceived as?[3]

The answer is that there is no way in which we can be said to know these things. The doctrine of structure was offered by both Russell and Eddington to support their causal theories of perception and their belief in an external world independent of the observer; but the analysis of Eddington's and Russell's concept of reality[4] will reveal that this belief is one of the basic axioms of their systems. We may object to assuming as an axiom such a fundamental and pivotal concept, but we would be wrong to suggest that Eddington was merely confused when he offered his

[1] *The Nature of Metaphysical Thinking*, p. 56.
[2] *Philosophy and the Physicists*, p. 134.
[3] *Ibid.*, p. 135.
[4] See chapter VI and VII of this study.

doctrine of structure as a way of supporting his realist concept of reality. Eddington's system should first be viewed from the point of view of its own internal logic; then, if we find sufficient reason, we can reject the axioms upon which it is based. 'Symbol' as applied to the sensible level depends for its meaning upon the recognition of the realist concept of reality as a basic axiom, but even this recognition does not render the use of 'symbol' perfectly clear.

It is the concept of mathematics as a realm of symbols which seems to have been Eddington's impetus for calling our physical knowledge 'symbolic,' but he clearly meant to apply the term to sense-data as well. Since the pointer-readings which constitute the data in science are formulated in mathematical terms, 'symbol' is seen to apply mainly to mathematics. The two-fold drive of relating mathematics to the physical world it is supposed to describe and of bringing the whole of exact scientific knowledge into some sort of harmony with the fundamental belief of common sense in an external world is reflected in the interest in mathematics as a symbolic operation. The meaning of 'structure' almost becomes identified with the mathematical symbol. The mathematician is called the "professional wielder of symbols" and praised as the proper person to be called upon to formulate our scientific knowledge of the world. (NPW, p. 209) It is the mathematical theory of groups which Eddington elsewhere describes as the scientific concept of structure.

> A terminable set of operations, or as it is technically called a *group*, has a structure which can be described mathematically. The fact that the operation which changes P into Q is always another member R of the group furnishes a set of triangular connections as the groundwork of the structure. These triangular connections can interlace a great variety of patterns; and it is the pattern of the interlacing which constitutes the abstract structure. (PPS., p. 140)

Mathematics is employed in science as a means of describing the group structure of the elements of our knowledge.[1] Structure

---

[1] Emphasizing the necessity of shifting our attention from 'things' to 'structures of relationship,' Martin Johnson (in *Science and the Meanings of Truth*) finds in the theory of groups the ideal of the meaning of 'structure.' By giving up all "pretence to know what the electron is and what it is doing, we can free ourselves to embody in matrix calculus or operator calculus the transformations which exhibit a structure. Then from the equations in that structure we deduce predictions verifiable in such concrete measurements as the displacement of a streak on a photographic plate." (p. 73) But the prediction or inference we wish

in the mathematical sense is something abstract but an abstract-
ness which somehow symbolizes a world outside mathematical
formulations. A group is defined as "a set of operators such that
the product of any two of them always gives an operator belonging to the set." (NPS, p. 262)[1] The product of physical analysis
of the atom, e.g., is "the *structure of a set of operations*." (*Ibid.*)
He has not forgotten his operationalism, since it reappears in a
modified form in the doctrine of structure where operations and
their interconnections define structure. But he never surrenders
his faith that this structure is a reflextion or symbolization of
something independent of and distinct from our operations.
Mathematics "dismisses the individual elements by assigning to
them symbols, leaving it to non-mathematical thought to express
the knowledge, if any, that we may have of what the symbols
stand for." (PPS, pp. 141–42) The mathematical concept of
structure is an abstraction from the more general concept which
we form of the entities involved in science. Infinite Euclidean
space, for example, can form the basis for an abstraction formulated in mathematical terms. The same is true of spherical space.

> Any point in spherical space can be changed into any other point by a
> rotation of the sphere. Thus to the points or elements of spherical space,
> A. B, C... there correspond operators P, Q, R, ... which are the rotations
> of the sphere; and the group of the operators is simply the group of rota-
> tions in the proper number of dimensions (in this case four dimensions).
> Regarding 'space' as a structural concept, *all* that we know about spherical
> space is that it has the group-structure of this group of rotations. When
> we introduce spherical space in physics we refer to something – we know
> not what – which has this structure. (PPS, 145–46)

The same is true, Eddington argues, of all the other mathematical constructs of the exact sciences: the construct is a
structural representation of an event or object beyond the realm
of observation. It is clearly only a conventional sign, it does not
describe but only names, although Eddington tries to convince us

to make is from equation to physical cause beyond the observable streak. Johnson does not recognize this crucial aspect of the doctrine of structure in Eddington's philosophy.
  [1] Like Eddington, J. L. Destouches has argued that physical reality is to be
contrasted with the reality of every day in that the former is not immediately
given but is an abstract construct inseparable from our tools of measurement.
The abstract character of physical reality finds its formulation for Destouches
in the mathematical theory of groups also. Cf. his *Essai sur la Forme Générale des
Théories Physiques*, pp. 26, 92.

that the mathematical manipulator can read off the name (his equation) and formulate an abstract, formal description of the events and objects supposed to be referred to by science. Physical knowledge is restricted to the structural concept, but it is always emphasized that this concept has reference to an objective reality. In this way, the mathematical constructs and formulations are related to the world believed in by common sense, although it is by no means clear what the details of this relation are. Moreover, Eddington's general doctrine of structure is weakened by the fact that order, in the mathematical development of the doctrine, ceases to be the defining characteristic of structure. The important mark of this doctrine is the ability of the mathematical theory of groups to scan a series or collection and extract in symbols and formulate a sufficient number of signs to form a consistent system.[1] The important epistemological aspect of the doctrine is submerged by this emphasis upon mathematics as the basis of our structural knowledge; now the one-one relationship between the two terms of the relata is difficult to envisage. For how can we say that for every feature in our mathematical formulae there corresponds a feature in the objective world? Mathematicians find it possible to use their equations as signs or symbols, but it would be difficult to argue for a causal relationship between formula and physical event. Nowhere is the double motivation behind Eddington's version of the doctrine of structure so clear as in his analysis of that doctrine on the basis of the mathematical theory of groups, for he there seeks to couple this theory of groups with the analysis of structure in terms of sensations and their similar physical correlates. Common sense has to be saved, but mathematical formulations must be given some objective meaning. The doctrine of structure seeks to accomplish both tasks, and in doing so it reveals the double nature of the concept of structure. Structure is related to mathematics via the theory of groups. The theory of groups is extended to non-mathematical collections, such as sensations and unsensed physical objects, while at the same time structure is related to the beliefs of common sense via the causal theory of perception and

---

[1] Cf. Krasner's "Une généralisation de la notion de corps," *Journal de math.*, 1938, 1er t., for a discussion and technical elaboration of the mathematical theory of groups.

the doctrine of separable causal chains. In terms of Eddington's
general philosophical position this latter connection of the doctrine of structure is the more important, but it is also that aspect
of the doctrine which is most in doubt and most vague.

The meaning of 'structure' is clearly not the same in its mathematical as in its sensory application. Suzanne Langer has called
attention to the symbolic character of mathematical knowledge
by saying that the mathematician

> does not profess to say anything about the existence, reality, or efficacy
> of *things* at all. His concern is the possibility of *symbolizing things*, and of
> symbolizing the relations into which they might enter with each other. ...
> Mathematical constructions are only symbols; they have meaning in
> terms of relationships, not of substance; something in reality answers to
> them, but they are not supposed to be items in that reality.[1]

But mathematical equations are symbols in a different sense
from sense-data, for they are able to be read as symbols only
because the scientist or mathematician attaches meanings to
them. They symbolize what we desire them to symbolize and
the symbol is substituted for and is a condensation of much more
complex concepts and processes. We can establish the symbolic
character of mathematical equations in virtue of the fact that
we attach to them specific facts; but, although sense-data are
claimed by Eddington and Russell to be symbols in the same
way, it is clear that this claim remains hypothetical. Sense-data
do not appear to resemble mathematical formulations; and,
although it is possible that, were we in posession of the key, we
could read off the physical causes of percepts as readily from the
percepts as we now read off the meaning or application of mathematics from the equation, it is hard to see how percepts and sense-
data can function as mathematical equations. There is an essential distinction between equations as symbols and percepts as
symbols, a distinction which Eddington seemed neither to
recognize nor to analyze.

Emmet suggests that the doctrine of structure is "an assumption which is a ghost (in terms of electro-dynamic analogies) of
the old 'copy' theory of correspondence."[2] I have suggested that
it is also another variant of the scholastic doctrine of form. In

---

[1] *Philosophy in a New Key*, p. 19.
[2] *The Nature of Metaphysical Thinking*, p. 59.

both cases, the doctrine is largely hypothesis, designed to save the belief of common sense in naive realism. After science has shown naive realism to be false, Eddington and Russell seek to patch up the theory by means of the doctrine of structure. We can no longer be said to perceive physical objects directly nor can we say that sense-data are parts of the surfaces of physical objects, but we can say we have some sort of intimate knowledge of physical objects in sensation. Eddington and Russell allow the sciences to take away the substance of naive realism but they reintroduce its shadow in the form of structure and symbol. In seeking to give the shadow some substance they are caught in two difficulties, either of which badly weakens the general doctrine. For the doctrine of structure to be able to harmonize science and common sense, Eddington and Russell have to argue for a one-one correspondence or isomorphism where structure is interpreted in terms of 'order'; but such isomorphism is very difficult if not impossible to establish. On the other hand, in order to bring mathematics into the merger of common sense and science, the interpretation of structure as isomorphism has to be abandoned. We have to conclude, therefore, that despite the elaborate arguments, the doctrine of structure as applied to the ontological problem of making inferences from sense-data to physical objects, is at best an hypothesis which if true would help to account for the inferences and would save the common sense belief in some form of naive realism. It should be recognized, however, that denial of the doctrine of structure does not necessitate the denial of the causal theory of perception, since the latter could be true without there being a relation of isomorphism between sense-data and physical objects. The two doctrines are related, but proof or disproof of one does not entail proof or disproof of the other. There is an essential discontinuity between the causal theory of perception and the mathematical or constructivist side of the doctrine of structure, a discontinuity which Russell has made more continuous by giving up the language of construction so prevalent in his *Analysis of Matter* and returning, in his latest work, to the language of inference of his earlier *Problems of Philosophy*. Eddington, on the other hand, has heightened this discontinuity by moving, in his last book, *The Philosophy of Physical Science*, towards a selective subjectivist position which, with its insistence upon *a priori* elements in our knowledge, would seem to demand the abandonment of the causal theory of perception in favor of the language of construction. Just as there is an intimate connection between the sensory side of the doctrine of structure and that of the causal theory of perception, so there is a similar close connection between the mathematical side of that doctrine and his claims for *a priori* knowledge. We must now examine this much criticized claim for *a priori* knowledge and see to what extent it can be made consistent with the doctrine of structure.

# CHAPTER IV

# SCIENTIFIC EPISTEMOLOGY

Eddington was not unmindful of the charge that there is no 'philosophy of science' but only particular philosophies of certain scientists; but he insisted that there is a single body of material in present-day science which does afford the basis for a philosophy of science.

> It is the philosophy to which those who follow the accepted practice of science stand committed by their *practice*. It is implicit in the methods by which they advance science, sometimes without fully understanding why they employ them, and in the procedure which they accept as giving assurance of truth, often without examining what kind of assurance it can give. (PPS, p. vii).

The task of the philosophy of science for Eddington is the investigation and explication of the methods and procedures of science. It is a problem of epistemology. Just as modern relativity theory induced a swing away from the older absolute point of view of reality to the operationalist attitude, so relativity theory had led scientists to a recognition of the role of their own techniques and methods in physical knowledge. The modern scientist has been forced to become a philosopher in the sense that he has found it necessary to consider epistemological problems.[1] The operationalism which has been accepted by many physicists and philosophers of science is a clear expression of this new orientation. Eddington believed operationalism was not the complete formulation of the epistemology presupposed by modern physics. The doctrine of structure examined in the previous chapter is another epistemological feature. A third and perhaps the most important

---

[1] In a searching review of the Einstein volume in *The Library of Living Philosophers* (7th vol. of the series; Evanston, Illinois, 1949), G. J. Whitrow (in BJPS, vol. II, 5) called attention to the epistemological nature of Einstein's reluctance to accept modern quantum theory. "Newton rejected the wave-theory because it seemed to him incapable of explaining all the known facts. Einstein rejects quantum mechanics because it appears to him to be diametrically opposed to his conception of explanation itself. His difficulty is epistemological and springs from his philosophy of science." (p. 61).

feature from Eddington's point of view is his claim for *a priori* knowledge. He does not use the language of the *a priori* until his last work, *The Philosophy of Physical Science*, but the basis of this claim can be found in his earlier books. Operationalism stipulates an observationally oriented science; the doctrine of structure seeks to show how observational knowledge can be applied to a world independent of observation, while the doctrine of *a priori* knowledge claims that much observationally derived knowledge can be predicted or inferred from an examination of the methodological tools of the physicist, together with an analysis of certain pervasive categories of thought. Operationalism and *a priorism* are the two extremes of Eddington's interpretation of the epistemological foundations of the exact sciences. The *a priorism* is historically later in Eddington's thought and marks the more essential aspect of his scientific epistemology. Always concerned with mathematics and its application to the sciences, he found in this field the ideal of the doctrine of structure as well as the impetus for the rational deductive side of his thinking. Vitally interested in the unification of modern science (his own work towards bringing about a harmonization between quantum physics and astronomical physics is noteworthy in this connection), Eddington saw in the *a priori* possibilities of knowledge a further means of reducing the hypotheses of science and of bringing about its unification. In contemporary science, Milne and Eddington have been the center of much controversy over their rigid claims for *a priori* knowledge, but Eddington has been alone, I believe, in attempting to work out the epistemology implied by this claim. He called this aspect of his philosophical interpretation of the sciences 'scientific epistemology', since it places the study of the epistemological presuppositions of science at the center of the understanding of the sciences.

One of the main virtues of an epistemology conducted within the physical sciences is, Eddington argued, that the epistemological conclusions can thereby be submitted to the "same kind of observational control as physical hypotheses." (PPS, p. 5) This is, however, a questionable assertion, since Eddington rests his case for *a priori* knowledge in part upon certain basic categories inherent in human nature, categories which it would not seem possible to alter under the impetus of observational stipulations.

But Eddington's case for *a priori* knowledge has another foundation, that of the accepted procedures in the sciences, which does permit of change and alteration. The test or indication that something is wrong with the methodological conventions of the sciences comes when they "lead to an *impasse* in the scientific developments." Eddington did not, however, pursue this point in any detail. It is offered in the spirit of getting his epistemological doctrines accepted by giving to them the sanction of science. The only example he offers of this observationally checked epistemology, of instances where observation has actually checked epistemological principles, is that observation can reveal fallacies in arguments or unwarranted assumptions (PPS, p. 5); but fallacies in arguments could certainly be disclosed by close analysis of the arguments, and unwarranted assumptions Eddington later claims to be discoverable by an analysis of the methods prescribed for measuring the entities involved. Thus, there seems little validity in his claim for an observational check upon the epistemology of the sciences. Scientific epistemology is not scientific in virtue of its kinship to empirical verification. Indeed, Dingle is correct in calling attention to the logical error in this claim, for "if we start with knowledge we cannot talk of *means of obtaining* knowledge, for that presupposes an *antecedent* equipment."[1] Eddington was too much taken with his newly developed system and with what he took to be its *a priori* possibilities and hence overstated the merits of his epistemological claims. His epistemology is scientific in no other sense than that it is the epistemology presupposed by the sciences. As such, its main task is to watch the observations, the procedures and limitations of the equipment of the scientist and to point out these limitations before investigators seek to use the methods. Specifically, the scientific epistemologist has as his task to examine the "procedure of good observation," the qualification 'good' being another unhappy label since all it is meant to convey is that the procedures from which the epistemologist attempts to predict the results be standard procedures, as specified by the sciences.

---

[1] *The Observatory*, vol. 63, p. 19. Dingle's review of *The Philosophy of Physical Science*.

Whether an observation is good or bad depends on what it professes to represent. A bad determination of the melting-point of sulphur may be an excellent determination of the melting-point of a mixture of sulphur and dirt. The terms used to describe an observation – to state what it is an observation of – imply by their definition a standard procedure to be followed in making it; the observer professes to follow this procedure, or a procedure which he takes the liberty of substituting for it in the belief that it will assuredly give the same result. (PPS, pp. 22–23)

The epistemologist has to "pick out the good observers – those whose activities follow a conventional plan of procedure." (*Ibid*, p. 23) A conventional or standard procedure is defined as follows:

The standard specification of the procedure of observing must be sufficiently detailed to secure a unique result of the observation. It is the duty of the observer to secure that all attendant circumstances which can affect the result, e.g., temperature, absence of magnetic field, etc., are in accordance with specification. Epistemological laws governing the results of the observation are such as are inferable solely from the fact that the procedure was as specified. (PPS, p. 20, n.)

It is this plan in which he was interested, for from it he believed the epistomogist is able to predict much of the knowledge actually obtained by applying the specified method.

The traditional method of systematic examination of the data furnished by observation is not the only way of reaching the generalizations valued in physical science. Some at least of these genererealizations can also be found by examining the sensory and intellectual equipment used in observation. (*Ibid*., p. 18)

Whereas the older physics sought to investigate an entity of a supposed external world, the modern physicist investigates knowledge, its limitations and possibilities. In this way, Eddington believed that

the whole system of fundamental hypotheses can be replaced by epistomological principles. Or, to put it equivalently, all the laws of nature that are usually classed as fundamental can be foreseseen wholly from epistemological considerations. They correspond to *a priori* knowledge, and are therefore *wholly subjective*. (*Ibid*., pp. 56–57)

It is important to note this initial qualification of his claims to predict *a priori* only the fundamental laws and not the special facts of nature.

The special facts, which distinguish the actual universe from all other possible universes obeying the same laws, are not given once for all at some past epoch, but are being born continually as the universe follows its unpredictable course. (*Ibid*., p. 64)

If we succeed at some time in predicting *a priori* some supposed fact, we should, Eddington recommends, alter our opinion and cease to call it a special fact. By definition, in other words, the special facts are accidental and cannot be predicted in advance of observation. Observation is itself necessary before *a priori* predictions can be made, observations which familiarize us with the meaning of the terms used and with the general nature of experience. Given this bare empirical acquaintance, Eddington believed that not only the general laws of nature but the cosmic constant can be predicted purely by an investigation of the methods employed in the sciences for measuring this constant. However, the result does not apply to the objective universe but only to the universe taken as the theme of physical knowledge. "The re-introduction of *a priori* physical knowledge is justified by the discovery that the universe which physical science describes is partially subjective." (*Ibid.*, p. 26) In reverse, "whatever is accounted for epistemologically is *ipso facto* subjective; it is demolished as part of the objective world." (*Ibid.*, p. 59)[1] Modern relativity theory has shown the amount the observer and his techniques play in formulating our scientific knowledge. It is this relativity which has led Eddington to insist that part at least of this knowledge, precisely that part which is dependent upon our techniques of measurement, is predictable from an analysis of those techniques. It is this aspect of scientific knowledge which he called 'selective subjectivism.'

In thus claiming to be able to make *a priori* deductions of some of the most important generalizations in science, Eddington, like Milne, emphasized the deductive, theoretical side of science.

Thus, whereas in the pioneer investigations of Einstein and de Sitter, Friedman and Lemaître, cosmology is derived from General Relativity as an extrapolation, in the more recent work of the two great British theoreticians, world-principles are axiomatic, and principles appropriate to local phenomena, e.g., General Relativity and other methods of analyzing gravitating systems, are, or should be, derived therefrom.[2]

---

[1] Cf. Eddington's letter in *Nature*, vol. 139, p. 1000 (his reply to Dingle's attack in "Modern Aristotelianism," *Nature*, vol. 139, pp. 784–86) where he expresses quite plainly this dependence of *a priori* prediction on the subjectivity of our physical knowledge. He also draws the distinction between laws which are predictable and special facts which are not.

[2] Whitrow, *The Structure of the Universe*, p. 92.

The derivation of world principles, of the general laws and constants of science, is made from two main sources: from an analysis of the methods actually employed in deriving the results of science, and from certain basic categories of the mind which are at work both in the world of science and in the world of common sense. "Familiar apprehension is subject to the same necessities of thought as those which, by more systematic application, yield the scientific description of the universe." (PPS, p. 133) Eddington contended that rough measurements or estimations of distances or length by means of ordinary sense organs do not differ in nature from the complex scientific measurements. Similarly, he argued that the apprehension of the world and the interpretation of our sense-data are both controlled by the same set of categories. He did not deny that the most advanced theories of modern physics may be working with new forms of thought not found in common sense, but he indicated that the frames of thought he was considering were those which furnish "most of our current vocabulary." (PPS, p. 117) And these are "more or less the same as that which corresponds to familiar apprehension of things around us." (*Ibid.*) Whitrow has correctly drawn the parallel with Kant. "Despite their divergent points of view, Milne and Eddington agree in their Kantian emphasis on the function of the mind in arranging the maze of natural events according to the mind's own canons of order."[1] It is the canons or categories of the mind which for Eddington play a general role in the interpretation of science.

We regard the mind as demanding by its 'necessities of thought' certain qualities in the parts which make up the physical universe. The mind imposes its demands by refusing to admit any system of analysis which does not yield parts with the required qualities. The fundamental laws of physics are simply a mathematical formulation of the qualities of the parts into which our analysis has divided the universe; and it has been our contention that they are all imposed by the human mind in this way, and are therefore wholly subjective. (PPS, p. 131)

It is through these categories that the final return of science to

---

[1] *Op. cit.*, p. 158. Eddington admitted his Kantian indebtedness. Vide PPS, pp. 188–89. "But if it were necessary to choose a leader from among the older philosophers, there can be no doubt that our choice would be Kant. We do not accept the Kantian label; but as a matter of acknowledgment, it is right to say that Kant anticipated to a remarkable extent the ideas to which we are now being impelled by the modern developments of physics."

common sense is to be achieved or understood. But the *a priori* deductions which can be made from these general categories are not the same as those which Eddington claimed can be derived from an analysis of the prescribed methods of science. Nor are they as controversial. The distinction between method and categories as a source for the *a priori* is an important distinction for understanding what is being claimed and for evaluating the claim. The distinction corresponds roughly to that between the sensory and intellectual equipment to which Eddington refers frequently in *The Philosophy of Physical Science*.

In introducing subjective selection (p. 16), I attributed it to 'the sensory and intellectual equipment' used in obtaining observational knowledge. The inclusion of *intellectual* equipment may have seemed surprising. It is easy to see that our sensory equipment has a selective effect – that the nature and extent of our knowledge of an external world must be largely conditioned by its lines of communication with consciousness, provided by our sense organs. It is not so obvious that within the mind there is any further selection at work on the material thrust upon it by the sense organs. (p. 114)

Sensory selection is broader than the selection introduced by our scientific procedures, but Eddington nowhere makes the claim that from a knowledge of physiology we can deduce *a priori* the kind of world we will perceive, although he would of course agree that physiology gives us specific evidence concerning the kind of sensations we are capable of having. His doctrine of *a priori* knowledge, however, is not offered on this level. He was concerned with the selective activities imposed upon the world through the prescribed methods of science and through what he took to be categories of the mind. He was not unaware of the fact that certain supposed categories, e.g., substance, find confirmation or exemplification on the sensory level; but he restricted the term 'category' to mind-contributed factors: they are predispositions inseparable from consciousness. (PPS, p. 132 ff.) What sensation does is to present us with "a world conformable to the mind's requirement of permanence." Our sensory equipment does make selections, but "this selection is altogether outside our present control." It is conditioned, Eddington argued, "by the fact that life would be impossible without some degree of harmony between the results of the selection and our engrained forms of thought." Like Kant, Eddington recognized the selective character of our sensory apparatus, although he would not name space and time as factors responsible to the sensory equipment; but both Eddington and Kant restricted the term 'category' to mental dispositions, setting them off from the sensibility. The *a priori* deductions stem not from the selective activity of our senses but from the frames of thought or the methods of scientific analysis. It was Eddington's firm conviction that where deductions can be made from either of these two sources (but chiefly, from the methods of science) these conclusions shed their problematic empirical character and take on a universality and a necessity which render the laws of science certain. (PPS, p. 20) The laws derivable *a priori* from these sources would seem, then, to become analytic. Given the kind of procedures we employ, and given the fact that we cannot apply our conclusions to an objective world, it follows of necessity that certain laws and principles are entailed logically in the system such that their denial is a self-contradiction. But Eddington was reluctant to apply this analytical quality to the general interpretations controlled by the mental categories. It is clear too that the kind of evidence which would confirm the *a priori* claim relating to knowledge derivable from the methods of science is not of the same nature or weight as that which would confirm the category claim. If the deductions which Eddington claimed from methods can be validated, his claim will then have some weight, but it is doubtful whether we could ever complete deductions from the supposed categories. The category claim is of the nature of an interpretation of our knowledge and must be judged in terms of its adequacy at encompassing the various facets of science. The claim for *a priori* knowledge derivable from methods is not an interpretation but is itself a method purporting to yield knowledge. Moreover, the methods employed by science can be and are changed, while the categories, although not considered as permanent and utterly unchangeable, are of a much more lasting nature.[1] They are more ingrained in human nature. It would fol-

---

[1] Eddington recognized that modern physics has freed us from many frames of thought, has discarded many potential categories; but he suggested that we may now be at the point where we can discover the ultimate categories which cannot be eradicated. "We are suspicious of the phrase 'necessities of thought';

low that those aspects of our knowledge which are controlled by these categories would have a universality; but, because of the speculative and interpretative character of these categories, they do not have the analytical necessity of those laws derivable from the specific method of science.

The categories are responsible for what Eddington calls 'frames of thought,' a predetermined disposition of the mind. One such frame of thought "is that which formulates the knowledge acquired by observation as a description of a world." (PPS, p. 115) Admitting the vagueness of the term 'existence,' Eddington argues that "It is a primitive form of thought that things either exist or do not exist; and the concept of a category of things possessing existence results from forcing our knowledge into a corresponding frame of thought." (Ibid., p. 155. Cf. p. 157) In fact, "so prevalent is this form that knowledge which is not concerned with the relatedness of sensory perceptions is often forced into it, and treated as a descriptive fact about a non-material world – a spiritual world." (*Ibid.*, p. 115) Another frame of thought is what Eddington termed the 'concept of analysis,' "that disposition of the mind to think of a 'part' in context with other parts, as a member of a system." (*Ibid.*, pp. 118–19) A third basic category is that of the 'concept of identical structural units,' the disposition which leads us to believe that the system being analyzed "is analyzable into parts which resemble" each other.

All variety in the world, all that is observable, comes from the variety of relations between entities. Therefore, when we reach the consideration of the intrinsic nature or structure of the entities that are related, there is nothing left but sameness. (*Ibid.*, p. 122).

The disposition at work here is "a habit of thought which regards variety always as a challenge to further analysis; so that the *ultimate* end-product of analysis can only be sameness." In other words, "the sameness of the ultimate entities of the physical universe is a foreseeable consequence of forcing our knowledge into this form of thought." To emphasize the subjective feature at work here, Eddington argued that,

for scientific thought has grown accustomed to doing without many of its alleged necessities. But, whether necessities or not, the forms we are about to discuss have a hold on us which seems incomparably stronger than any we have hitherto thrown off." (PPS, p. 118).

There is nothing in the external world which dictates this analysis into similar units, just as there is nothing in the irregular vibrations of white light which dictates our analysis of it into monochromatic wave trains. The dictation comes from our own way of thought which will not accept as final any other form of solution of the problem presented by sensory experience. (*Ibid.*, p. 125)

Although the identical units are conceived as being related in a system of similar parts, a fourth category is distinguished by Eddington, the 'concept of interaction,' a concept which seeks to conjoin the identical units of a system. This category refers specifically to quantum phenomena, where it is now an accepted fact that the tools of investigation interfere with the objects being observed. Thus, an objective interaction, even if it existed, would be unknown. Interaction must be introduced from the side of the subject. The deviation of a particle from its course, for example, can be due either to an actual force acting upon it "or because, owing to the observational indistinguishability, another particle has been mistaken for it." (*Ibid.*, p. 128)

Eddington's discussion of this supposed category of interaction is far from satisfactory. The use of mental categories as interpretative factors for the actual findings of physics comes to the fore in his description of this category. If there is any point at which Eddington's claim for *a priori* knowledge is forced, it is in this supposed concept of interaction. He has cleverly blended the known subjective factors of quantum physics (the indistinguishability of particles) into the category of interaction present in all men, at work on scientific as well as on the common sense level. He may have succeeded in describing the actual workings of the minds of scientists as they were confronted with the results of quantum physics, but he seems to have confused explanation within the sciences with a controlling category. It is doubtful whether his entire case for categories can be substantiated, although it must be recognized that his claims for the categories of existence and ultimate identical units do have a degree of plausibility. It may be that the interactions we think we observe in objects outside us are really subjective interactions induced by the knowing subject's means of observation. In fact, "there is now strong reason to believe that *all* interaction forces in physics arise from the indistinguishability of the ultimate particles." (PPS, p. 128) But even if true, these facts do not convert the interaction into a category. The relation between relativity physics, with its emphasis upon the reference frame as a fundamental ingredient in our knowledge of the world, and Eddington's claim for *a priori* knowledge is intimate. Eddington is able to advance his claim because the physical knowledge of modern physics is admittedly infused with a large element of subjectivity or relativity. But we must keep constantly in mind that the *a priori* for him stems from two sources: the methods of science and basic categories of the human mind. It is much more likely, I believe, that a case can be made for interaction being the result of methods of quantum investigation than that it is a category. Eddington does not always keep the distinction between the two sources of *a priori* knowledge clear, and here is one point at which he has blurred this important distinction.

He is on much safer ground when he argues that, like existence, substance is "one of the most dominant concepts in our familiar outlook on the world of sensory experience." (PPS, p. 129) Permanence is the usual way in which this category enters into our scientific formulations, conservation of mass, of energy, of momentum. Thus, the identical units which we seek, under the control of the category of analysis, are conceived as being indestructible. Substance is a category which appears early in Eddington's writings. He himself tells us that it was the first area in which he suspected a selective subjective factor in our knowledge. (PPS, p. 130) In *The Nature of the Physical World*, he insisted that the permanence in our physical knowledge is due to the "mind having demanded permanence. ... The element of permanence in the physical world, which is familiarly represented by the conception of substance, is essentially a contribution of the mind to the plan of building or selection." (p. 241) He even speaks of this category as an "innate hunger for permanence." However, in *Space, Time, and Gravitation*, the view of substance as a category was not formulated, although he credited the appearance of permanence in physical knowledge to the activity of the mind. But in this early work, the selective work of the mind in this connection was the result of its insistence upon "regarding only the things that are permanent" as important and basing its formulations upon these. (p. 197) Permanence enters the world of *Space, Time and Gravitation* not by an ingrained disposition of the mind which imposes permanence on the world but in virtue of the preference the mind has for permanent events. Permanence enters the world of *The Philosophy of Physical Science* via two routes: through the ingrained categories of the mind and through sensation which mysteriously mimics the demands of the mind. The controlling force of this category lies in its ability to force us to continue analysis until we have reached permanent or semi-permanent combinations of parts and characteristics which remain permanent in the vicissitudes of phenomena. (PPS, p. 136) Eddington does not say even in the last and most radical formulation of the doctrine of categories that the mind imposes permanence upon the world, although such an assertion is implicit in his selective subjectivism. But the world so controlled is the subjective world described in physical knowledge: it is not the objective world of common sense or the active world of the causal theory of perception. However, it is important to notice that even within this subjective world, the category of substance (in many ways the most important category for Eddington) does not yield specific results, but only controls the general outlook of the scientist. No specific *a priori* deduction is claimed to stem from ingrained dispositions, although in many passages Eddington writes as if he were about to produce such a deduction. All that he really asserts is that the world appearing in the descriptions of the physical sciences will show elements of permanence. The law of conservation fits in with the demands of the mind for such constancy, but this law itself is not deducible from a knowledge of the disposition of the mind towards permanence. Even though Eddington fails to disentangle the two basic sources of *a priori* knowledge in his claim, arguing that "by consideration of certain deeply rooted forms of thought we can foresee the fundamental laws and constants which occur in the physical description of the universe" (PPS, p. 134), he nowhere makes or indicates that it is possible to make a deduction of these laws and constants from a knowledge of the categories. At most, from the categories discussed by Eddington, we could predict that the world described in physics would consist of more or less permanent, discrete entities related together in a system. But it is clear that even this much could not be predicted in advance of experiments and prior to the development of scientific descriptions, since the conviction that the mind has these categories or dispositions provides no basis for inferring that these will appear in the final description. Eddington has not shown, as Kant thought he had, that the world actually known and described would be impossible were it not for the existence of certain basic categories. Such a failure on his part need not, however, bother us, for it is not to his doctrine of categories that he appeals in his crucial and controversial claims for *a priori* knowledge.

All of the laws and constants claimed to be deducible *a priori* are said to be deducible from a knowledge and analysis of the methods employed by the sciences. Even here there are no clear-cut deductions. Whether we wish to accept the doctrine of categories, in either Eddington's or Kant's version, makes no difference to the issue of *a priori* knowledge in Eddington's interpretation of the sciences, for it is not to these that the asserted *a priori* knowledge is traced. At most, the categories influence our selections, give us a disposition towards one sort of analysis rather than another. They are not, as they were for Kant, the presuppositions of the possibility of experience. No transcendental derivation of the categories is suggested and no logical deduction of the laws and constants is made from the basic categories. In brief, Eddington's doctrine of categories offers a suggestive account of certain dominant traits of our knowledge, but it provides little specific detail for his general *a priori* theory. The *a priori* of procedures and not of categories constitutes the most important and concrete part of his doctrine.

In placing the main weight of the *a priori* claim upon the methodological analysis of scientific epistemology, Eddington attempts to show that the end result of many parts of physical knowledge has already been assumed prior to the operations of science through the methods stipulated for investigation. He often states that "where science has progressed the farthest, the mind has but regained from nature that which the mind has put into nature." (STG, pp. 200–01) The same book ends with the following elliptical remark:

We have found a strange foot-print on the shores of the unknown. We have devised profound theories, one after another, to account for its origin. At last, we have succeeded in reconstructing the creature that made the foot-print. And Lo, it is our own. (p. 201)

After completing his *Relativity Theory of Protons and Electrons*, he concludes with a similar reflection: "in the end what we comprehend about the universe is precisely that which we put into the universe to make it comprehensible." (p. 328) As early as 1936, in this same work, the general outlines of his *a priori* claim were set forth.

> We conclude with a brief reference to the philosophical position towards which the present results trend. Unless the structure of the nucleus has a surprise in store for us, the conclusions seem plain – there is nothing in the whole system of laws of physics that cannot be deduced unambiguously from epistemological considerations. An intelligence, unacquainted with our universe, but acquainted with the system of thought by which the human mind interprets to itself the content of its sensory experience, should be able to attain all the knowledge of physics that we have attained by experiment. He would not deduce the particular events and objects of our experience but he would deduce the generalizations we have based on them. (p. 327)

From a close analysis of microscopic and sub-atomic phenomena, Eddington drew the general conclusion that all laws of physics, even on the macroscopic level, involved the same relational connection with the observer and his techniques as admittedly operates on the atomic level, a connection which he thought yielded *a priori* predictions. It is true that the pecularities of atomic physics do lend support to a subjectivist thesis: there is an inseparable connection between results and the means of deriving the results. But many of Eddington's statements sound as if he meant to claim, even on the atomic level, an ability to predict *a priori* all the main laws of atomic physics. Such a claim is slightly exaggerated unless carefully qualified, but, when extended to the macroscopic level, the claim becomes either patently false or reduces to a truism. Already in 1936 he had made the jump from atomic to general physics, and it was only three years later that (in his Tarner lectures) he gave an extended exposition of the claim for *a priori* deductive knowledge in macroscopic physics. The difficulty, however, is to find specific arguments or demonstrations in Eddington's writings showing how the asserted deductions are to be carried out from an analysis of the methods of science. *The Philosophy of Physical Science* contains many statements about the virtue of scientific epistemology and its ability to arrive at knowledge *a priori*, but it is admittedly concerned with the test case of the cosmic constant more than it is

with the many other similar instances of asserted *a priori* knowledge. Even in the case of the cosmic constant we do not find
specific or detailed arguments proving his claim. But before we
consider the information and hints he does give regarding the
deduction of the number of particles in the universe, we must try
to piece together Eddington's various detached remarks concerning the purported *a priori* knowledge of specific laws of physics
derivable from a knowledge of scientific methods, and to see how
much of a connected account we can construct from these sources.
At the same time we must bear in mind that, although there is a
sharp separation between the two sources of *a priori* knowledge
in Eddington's account, the *a priori* categorial claim plays a
fundamental role in the general interpretation of physical science,
affording the basis for Eddington's final union of common sense
and science. But this aspect of the doctrine is more speculative,
while the claim for *a priori* knowledge from an analysis of the
methods of science presents a more plausible case and is argued
for as being more rigid and certain.

One reason for the difference in the tone of the two claims is,
as we have seen, the weight lent by the general relativity theory
to the *a priori* of method. Eddington frequently argued that the
Euclidean or non-Euclidean character of space is dependent
upon our processes of measuring space. Operationalism is the
formalized expression of this aspect of relativity theory. In other
passages, the Fitzgerald contraction is presented as a piece of
knowledge which can be determined either by experiments or
by analysis of the nature of measuring rods and the laws of
electromagnetism. (NPW, pp. 5, 6.) He writes, in this connection,
of the necessary consequence of the contraction from the known
scheme of electromagnetic laws, calling attention to the quantitative calculations supporting the results of experiments made by
Lorentz and Larmor in 1900. Elsewhere he argues that the
similarity of units in the groups analyzed by physics results from
the mathematical process of group analysis. The mathematician
is said to show

that those observational effects which reach our perceptions, generally
attributed to the fact that we are dealing with an assembly of like individuals, are deducible more directly from the fact that the assembly is
obtainable by a kind of operation which, once performed, can be repeated
any number of times without making any difference. (NPS, p. 265)

Such claims as these are clearly not controversial; for, all that they assert is that, given a knowledge of the nature of material or optical measuring appliances together with the laws explaining the behavior of their components, predictions can be made concerning the behavior of the appliances in certain specified conditions. Such predictions amount to no more than the ordinary predictions made from scientific theories and hypotheses, except that in these instances we find that our predictions possess a high degree of certainty, especially when quantitative calculations of results can be made in advance of the confirmatory experiments. Most if not all of the claims of *a priori* knowledge made by Eddington, on the basis of the methods of science, can be seen to reduce to just this recognized ability to calculate the results in advance of experiments, once we are given certain laws and other general information. Eddington was primarily interested in the mathematical formulations of science; it is with the exact sciences that he is concerned and to which he refers in his philosophical interpretation of science. The aim of mathematical deductions is, as he explains in *The Mathematical Theory of Relativity*, (1) "to examine how we may test the truth of the fundamental postulates" of science, and (2) "to discover how the laws which they express originate in the structure of the world." (p. 105) He there suggests that the ideal would be to work both ways, from experiment to mathematical formulations and from mathematical formulations to the results of experiment. The claim for an *a priori* of method was, I believe, no more than this ideal of an axiomatic deduction in mathematics of the major results of scientific experimentation. The fact that what Eddington meant is not readily recognized is due to Eddington's failure to see clearly what he himself meant. His choice of the term '*a priori*' for such knowledge has caused much of the confusion in his own mind and has led many critics astray, but that this is all that his claims amount to can be shown clearly in at least one of the two major claimants for this sort of *a priori* knowledge, that of Einstein's law of gravitation; the claims for *a priori* deduction of the number of particles in the universe, though very abstruse, seems to follow the same pattern.

Setting aside for the moment his frequent claims that the law of gravitation as formulated in general relativity is a result of

the mind's inherent desire for permanence, we find Eddington as
early as 1920 writing that conservation "is simply a mathematical
identity due to the way in which the expression has been built
up from the simpler elements, $g_{\mu\nu}$." It results from this property
that, *"provided we measure space and time in one of a certain
limited number of ways,* matter will be permanent."[1] He argues
that to predict the laws of conservation of matter, "we need to
know nothing about the properties of the constituents of the
external world; all that we need to know is, under what names will
mind recognize the things which obey the laws?"[2] That is, we
need to know only the way in which the mathematician will set
about calculating the results of our measurements. In the same
year, he argued that mass and momentum, two important factors for gravitation, are dependent upon (1) the four identical
relations between the ten quantities of matter, and that these are
in turn dependent upon (2) the way the $G_{\mu\nu}$'s are constructed by
definition, and (3) upon the way in which the mesh system is
drawn up for plotting the measured world. (STG, p. 195) The
truth of the law of gravitation "cannot be regarded as subsisting
apart from the experimental procedure by which we have as-
certained its truth."[3] Later, he asserted that the statements
"there is a universal law of gravitation" and "the second order
equation (in empty space) is $G_{\mu\nu} = \lambda g_{\mu\nu}$," are both "rigorously
deducible from an examination of the observational procedure
followed in obtaining the measurements which are deemed to
establish the law of gravitation." (PPS, p. 47) He admitted that
the second of these statements is not at present deducible directly
from the procedures used, since it depends upon another statement, "the law of gravitation is expressible by a differential
equation of the second order," and this statement waits upon
experience. He stressed the relation between the tensor $G_{\mu\nu}$,
the mesh system of measurement, and the resulting concept of
space, arguing that we can see from pure reflection that any
change in any one of these ingredients will call for a corresponding
change in the others. In thus pointing out this particular set of
relations, Eddington had in mind Einstein's remarks concerning

---

[1] "The Meaning of Matter", *Mind*, Oct. 1920, p. 153.
[2] *Ibid.*, p. 154.
[3] *Ibid.*

the way in which the fundamental tensor used in formulating the nature of the gravitational field in mathematical terms, determines at the same time the metrical characteristics of the four dimensional space of modern physics.[1] De Sitter also has stressed this same relation between the co-ordinate system used in measurement and the type and amount of gravitation. "We can thus produce or abolish a gravitational effect by a simple transformation of co-ordinates. ... There must thus be an intimate connection between the values of the coefficient $g_{ij}$ and the gravitational field."[2] Gravitation becomes a property of space "in this sense that the same coefficients of $g_{ij}$ which determine the metric properties of the four-dimensional time-space, also determine the gravitational field."[3] In other words, the postulate of general relativity theory – the "postulate that all laws of nature shall be invariant with regard to *any* transformation of co-ordinates – leads with logical necessity to a definite theory of gravitation."[4] The procedures of measurement determine the kind of space and the nature of the gravitational field. In the ideal of a deductive theory for which Eddington was always striving, this reciprocal relation between mesh system, space, and gravitation becomes a basic axiom, a definitional matter from which the laws of motion can be mathematically deduced. In *The Mathematical Theory of Relativity*, with this as an axiom together with four other axioms concerning (a) the interval as the primary unit of pairs of events, (b) the formulation of the interval in terms of the quadratic function of the difference between various coordinates, (c) the path of a freely falling particle forming a geodesic, and (d) the track of a light wave being a geodesic with ds = 0, Eddington produced a standard deductive formulation of the results of modern physics.

Modern science was radically re-oriented by the general theory of relativity. It was injected with a good portion of what Eddington called 'selective subjectivism.' Talk of discovering properties of an objective world was replaced by talk of a world inextricably connected with the observer and his tools of measure-

---

[1] "Die Grundlage der Allgemeinen Relativitätstheorie," in *Annalen der Physik*, vol. 49, 1916, p. 779.
[2] "Space, Time and Gravitation," *The Observatory*, vol. 39, p. 416.
[3] *Ibid.*, p. 418.
[4] *Ibid.*, p. 412.

ment. In terms of philosophical doctrines, modern science has swung away from the naive realist position said to be inherent in common sense (and adopted, with modifications, by the sense-datum philosophers), towards something like the phenomenological view of Husserl and Whitehead, where the subject cannot be separated from the object in an ontological sense; except of course that modern science has not as a rule been much concerned with the actual perceptual problem, the genesis of sensations or ideas. Reflection upon the implications of this new view of reality led Eddington to make extravagant claims for knowledge. He seems to have argued that if knowledge is partially subjective, to the extent that it is subjective it should also be predictable *a priori*. The final evaluation of this relation between subjectivity and a priority cannot be made until we have fully analyzed Eddington's concept of reality, but it takes little insight to see the extravagance of such statements as the following:

> I am almost inclined to attribute the whole responsibility for the laws of mechanics and gravitation to the mind, and deny the external world any share in them. ... All that nature was required to furnish is a four-dimensional aggregate of point-events ... For the use made of the point-events mind alone is responsible. [1]

It is not always clear what Eddington means by 'mind,' but, as used in his frequent claims for *a priori* knowledge, a mind-dependent event need not be also an event predictable *a priori*. Eddington might have agreed that his purported categories of the mind were *a priori* in the logical sense of being presuppositions of knowledge, but he could not have maintained this interpretation in his claims for a methodological *a priori*. Deductive derivation from methodological principles is the meaning he attached to this aspect of his claim. It is a separate problem to decide why such deductive derivations as are possible in mathematical physics are possible, why the chain of reasonings constructed by mathematicians can successfully mimic or reproduce the chain of events in nature. Mind certainly plays its part in making possible such exact formulations; but, from the fact that such deductions can be made from a few axioms, we are not entitled to infer that the resulting knowledge is predictable in advance of experimentation. He was not, as Whittaker has pointed

---

[1] "The Meaning of Matter," *Mind*, vol. 29 1920, p. 155.

out, advancing the thesis that a large part of our physical knowledge "can be derived by pure cogitation, without depending in
any way on the results of observation and experiment."[1] It is true
that in many places Eddington asserts that epistemological
principles are adequate to replace observation and experiment,
but Whittaker is correct in saying that "an examination of his
published work shows that in effect the epistemological principles
were by no means independent of knowledge derived from sense
perception."[2] The deductive side of Eddington's thought was
never abstracted from the experimental and inductive means of
obtaining knowledge. Eddington never did claim that it would
have been possible to deduce the present law of gravitation from
an analysis of our methods of measurement of space, prior to
Newton's empirical tests. He did not, that is, claim that starting
from scratch we could derive the law of gravitation by deductive
processes or that mind is imposing its own patterns upon reality
without any reference to past experience. He admitted that in the
act of constructing, the world to be built must follow the specifica-
tions of the common sense world. The deductions are made within
the context of a vast body of experimental data, but unless we
wish to give a new and rather strange definition to the term 'a
priori,' it would seem to be a misnomer to give this mathematically deduced knowledge such a label. A better case can be made for
the category of substance or permanence being *a priori* than there
is for calling the law of gravitation a law foreseeable *a priori* or
a law resting upon presuppositions of thought. It is important to
stress the intimate relation between the gravitational constant
and the metrics of space, and to call attention to the subjective
implications of this relation; but little is gained and much clarity
lost by describing this as an *a priori* relation.

The application of mathematics to the results of the physical
sciences has led many into believing in some peculiar kind of
relation between mathematics and reality. It is startling to see
the way in which mathematical formulations are able to stand
for symbols of reality in the sense of enabling the mathematician
to read off from the symbols the appropriate data and apply
them to reality, although in a certain sense the whole connection

[1] *From Euclid to Eddington*, p. 185.
[2] *Ibid.*

has been pre-arranged and decided upon. Thought in general, as symbol, evokes admiration in the mind of the philosopher, but these useful instances of the symbolization of reality do not form any basis for the extravagant claims advanced by Eddington for an *a priori* prediction from symbols to object. There is a special sort of relationship on the atomic level between the tools of investigation and the results of investigation and it was undoubtedly this area of science which gave the impetus to Eddington's wide formulation of his doctrine of scientific epistemology; but even in this more favorable area, there is no clear evidence that he had discovered some unknown property or a new mathematical formulation permitting man to make proper *a priori* predictions of reality. There is a good deal of truth in Eddington's exaggerations concerning the relational character of modern scientific knowledge. The possibility of a deductive formulation of most of science is one such important truth. The next chapter seeks to sift from his many obscure statements those portions of truth which do belong to modern quantum physics concerning the peculiar relation between tools and results, seeks that is, to place the claim for the deducibility of the constants of the universe in the general context of a deductive system. Even in this area of physics, we shall find that there is no mysterious *a priori* waiting to be seized which will replace all empirical investigations.

# CHAPTER V

# RATIONALISM AND EMPIRICISM IN MODERN SCIENCE

In his *Analysis of Matter*, Russell makes the following observation on the deductive character of modern science:

> The appearance of deducing actual phenomena from mathematics is delusive; what really happens is that the phenomena afford inductive verification of the general principles from which our mathematics starts. Every observed fact retains its full evidential value; but now it confirms not merely some particular law, but the general law from which the deductive system starts. (p. 88)

He points out that "the unity and simplicity of the deductive edifice ... must not blind us to the complexity of empirical physics, or to the logical independence of its various portions." (p. 89) He remarks that what is novel and interesting in Eddington's *The Mathematical Theory of Relativity* "is the character of the relation between empirical and deductive physics. But there is no real diminution of the need for empirical observation." (*Ibid.*) Despite the imposing deductive constructions made by modern physicists, the role of empirical investigation has not been reduced in importance. Among contemporary scientists, there are many who emphasize one side at the expense of the other, but few have been able to see the interconnection between rationalism and empiricism in scientific investigations and formulations. Milne and Eddington represent those who have stressed the rational side of science, a side which, judging from the reactions of their critics, was in need of emphasis. It is probably correct to say that modern science is moving more towards rationalism than towards a simple empiricism, but this, if true, is true only because the modern edifice of science has been so constructed that most of the relevant empirical data in many fields have been collected. The task that remains is to unify these data and to formulate them in a deductive system. When new data arrive, they can be incorporated into this system or it may be found that a new system of deductions will be required. The advance of physics is not improperly characterized as a movement from the collection of data, the formation of theories, the mathematical formulation of the data, the theories and laws, to the final formulation of these facts and theories into a rational, deductive system. At this point, it frequently happens that new discoveries or new theories necessitate the alterations of the established system. But the movement from induction and empiricism to deduction and rationalism is clear. Physics today seems to have entered upon this final stage of deductive formulation. For such formulation, J. L. Destouches[1] seems correct in insisting upon three important steps: the synthesis of data and current theories, the statement of the axioms to be used in the system, and the actual deductions in the system. Eddington followed this pattern in his early *Mathematical Theory of Relativity*, although not with the meticulous care shown by Destouches; but in his subsequent works, he moved towards the general *a priori* claims which the success of his deductive system had suggested and hence not only failed to comprehend that his claims for *a priori* knowledge were still claims for a rigid deductive system, but failed as well to provide detailed statements of his axioms, his rules, and his primary data. The result was general confusion, both in his own mind and in those of his critics. But one of the more valuable products of the confusion and controversy evoked by both Eddington's and Milne's claims was a critical examination by most scientists of the relation between rationalism and empiricism. Before we assess the validity of these critical examinations, placing Eddington's general claim in the context of these debates, we must make a few remarks about his claim that the number of constants in the universe can be predicted in advance of empirical discovery.

The claim here is much more obscure than the other claims for *a priori* knowledge and presupposes a unique kind of mathematical formulation which few admit of understanding. I cannot attempt to assess the validity of his mathematical deduction. Broad was wise in urging someone who is competent to make a careful check of Eddington's calculations,[2] since only then can a final evaluation be made of this aspect of his doctrine. Few have

---

[1] *Principes Fondamentaux de Physique*, t. 1, p. 10.
[2] Vide his review of PPS, *Philosophy*, vol. xv, 1940, 59.

undertaken this task. As Noel B. Slater says about the book which contains Eddington's calculations (*Fundamental Theory*), "Many have tried to read the book; but all have found it difficult."[1] It is this aspect of Eddington's philosophy which raised the storm of protest from his critics and which led many to re-examine the general relation between rationalism and empiricism. Physicists have either dismissed the claim as extravagant or absurd (e.g., Dingle), or, like McCrea, have confessed that the mathematics and the language of Eddington's formulation of the deduction is almost unintelligible.[2] A few, such as Burniston Brown,[3] Whitrow,[4] and Whittaker, who edited the posthumous *Fundamental Theory*,[5] have professed to understand and to follow the deductions; but their explanations are more concerned with showing how the deduction could be made than with an actual detailed explanation of the steps of the argument. Where the argument is explained, so far as I have been able to determine, all that has been done is to make a deduction from certain basic premises. But Whittaker claims that Eddington's calculations have been confirmed by diverse methods, urging that the new fundamental theory is "a doctrine original in its deepest foundations, a new isomorphism between Thought and Nature."[6] Its originality, however, needs to be checked by additional mathematicians who will take the claim seriously. The latest work on this aspect of Eddington's thought, that of Slater, is an attempt to collate and reproduce a good sample of the various drafts Eddington made of his *Fundamental Theory*. Slater's work ought to enable now a careful analysis of the deductions and calculations.[7] Previous to Slater's work, the time was not ripe for an open minded attitude towards Eddington's claims; the theory was received with wild emotion during Eddington's life time. One of the biggest factors causing

---

[1] *The Development and Meaning of Eddington's 'Fundamental Theory,'* Cambridge, 1957, p. 1.

[2] Vide his review of *Fundamental Theory*, *Math. Gaz.*, vol. 31.

[3] "Why do Archimedes and Eddington both get 1079 for the Total Number of Particles in the Universe?," *Philosophy*, vol. xv, 1940, 59.

[4] *The Structure of the Universe.*

[5] "Eddington's Theory of Constants of Nature," Math. Gaz., vol 29, pp. 137–44.

[6] *Ibid.*, p. 144.

[7] Vide the bibliography in Slater's book for a careful listing and brief analysis of the numerous articles and discussions relevant to Eddington's *Fundamental Theory*.

this reaction was the language in which Eddington stated his claim, for the label of *a priori* was, as he recognized, anathema in scientific circles, where experimentation was the ruling word. But if my contention is valid that the label is strangely misplaced in his general application to the *a priori* of method, we may be able to find reasons for taking the claim of a deduction of the constants of the universe more seriously. At most all that I can hope to do for the discussion of this issue is to indicate where the claim seems to follow the pattern of the claim advanced for the law of gravitation, fitting into a scheme of a mathematicization of science.

The contention that the constants of the universe can be quantitatively fixed through a mathematical chain of deductions working only, as Whittaker is fond of repeating, from qualitative premises, stands in a slightly different position from the contention that the law of gravitation can be deduced from a knowledge of our selected measuring system of space-time; but the two claims are on the same footing in that both recognize an intimate connection between the standards specified for obtaining the results and the results actually obtained. The former claim, unlike the latter, occurs within the context of quantum physics, where it is generally recognized that the connection between tools of measurement and results is much more intimate, a connection which Eddington largely ignores in the development of his *a priori* claim. Eddington points out that his claim concerning the deducibility of the constants of nature is not based upon any conviction that the number of electrons and protons can be actually counted, as we would count apples in a barrel. In fact, just the reverse is true: the claim is made because it is impossible to count electrons and protons in this simple and direct fashion; the method of estimating the number of entities in this area is indirect and inseparable from the result.

> When knowledge is claimed of the number of protons and electrons in a gram of hydrogen, the observational procedure referred to cannot be counting. It must be knowledge of the result of some other procedure by which an integral number is affixed to a system. (PPS, p. 172)

The calculation is made upon uncountable entities and is made possible by the peculiarities of quantum arithmetic. He speaks of the result of the calculation as being 'foisted on us by quantum theory' as a result of the method of numerical analysis of quantum phenomena. (PPS, p. 177) He is claiming, in other words, that the nature of the arithmetic employed in the quantitative calculations of quantum physics leads to definite quantitative results independent of any considerations of the nature of the phenomena being formulated. The indirect means of counting electrons and protons determines the numerical result of applying that means. Whittaker has argued that Eddington assumed much of qualitative physics as true, such as "the identity of mass and energy, the theory of the energy-tensors, and the interpretation of its elements, the exclusion principle," but refused to assume any number determined empirically.[1] Given the general data of physics and these qualititative principles, he has sought to establish a deductive system which can produce the quantitative figures, "the exact values of the pure numbers that are constants of science – the numbers that are analogous to the number $\pi$ in geometry."[2] Like the Pythagoreans, Eddington was strongly attached to numbers and saw in them a strange power for explaining the world. He was not so rash as to establish a religion or cult devoted to explaining all reality in terms of numbers, but in much the same fashion as Pythagoras, Eddington argued that a close analysis of numbers and their relation to non-numerical reality leads to valid insights concerning the structure of the world.[3] The world for Eddington's scientific epistemology is a subjective world, the world of observer and object; but he believed that numbers as well as categories of the mind play a fundamental rôle in determining the nature of this world. Moreover, he never forgot his dedication to the dualism necessitated by the modifications of naive realism. Hence, numbers stood as symbols not only of the phenomenal world which they help to form, but also of the non-phenomenal world revealed in our structural knowledge. The very fact that numerical measurement involves four entities,[4]

---

[1] Whittaker, op cit., pp. 137–138.

[2] *Ibid.*

[3] Brown has called attention to the similarity between Eddington and Archimedes on the number of constants, both men being vitally concerned to determine this number and both, oddly, arriving at the same figure. Op. cit., *Philosophy*, vol. xv, 59.

[4] "A measurement of length is a comparison of a relation of extension between two entities in the system under observation with a relation of extension between two entities marking the ends of an adopted standard." (PPS, p. 168).

leads, he believed, to the fact that the world of physics is four-dimensional.

From the association of measurement with four entities we are led, without further investigation, to expect that the number 4 will in some way make itself evident in the world-picture which embodies the results of our measurements. It is the seed from which spring those oddly assorted pure numbers which we call constants of nature. This conclusion in itself tells us very little, and gives no warrant for numerological speculations. I believe that the number 4 introduced in this way is actually responsible for the four dimensions of space-time, *but only indirectly*. In an actual calculation the number of dimensions of space-time is reached by the route $\frac{4.3}{1.2} - 1 - 1 = 4$ and it is a coincidence that the number we end with is the number we start with. (PPS, pp. 168–69).

In this way, Eddington found it possible "to calculate in spherical space the number of independent wave-functions of this form with the necessary relativistic property; and the total number of elementary particles corresponds to this number."[1]

If it is correct to say that the indirect means of measuring the phenomena in quantum physics is analogous to the methods used for determining the nature of the space-time continuum in macroscopic physics, in that both methods are so constructed as to play a determining rôle in the final product, then Eddington is entitled to claim an interesting and significant deduction. But his claim that the number four introduced by necessity into measurement leads to the four dimensional space-time continuum must be dissociated from the valid deductions that may be made within his system. This Pythagorean claim is on the same tenuous base as Eddington's general categorial claim: it is an interesting but wholly unsupported doctrine. That it is possible to take the qualitative principles of physics and the numerical quantities of Einstein's space-time continuum and from these deduce the number of constants in the universe, may be proved. But this would hardly present us with an *a priori* deduction: it would be no more than is commonly done in mathematical physics, starting with data and principles and deducing quantities. Moreover, Eddington's Pythagorean-like approach to his *a priori* doctrine leaves to one side the more interesting features of quantum physics, i.e., the suggestive inseparable role of the observer. He could have made a better case for his doctrine by drawing the

[1] Whittaker, "Eddington's Theory of Constants," *op. cit.*, p. 139.

parallels of the relation between measurement, on macro- and microscopic levels and the results of measurement. But Eddington was content to build his theory on the tenuous grounds of a Pythagorean conviction concerning the relation between numbers and reality. He was thus led to believe that the mathematical deductions he made within quantum physics were confirmation of his Pythagorean convictions and established an *a priori* union between numbers and reality.

There is an important subjective factor at work in the whole of quantum physics: this subjectivism has considerably altered the philosophical interpretation of the physical sciences. The contributions of mathematics to contemporary physics do not lie in any mysterious isomorphism between numbers and reality which enables us to make bold *a priori* predictions concerning reality, but in making it possible for us to deduce more definite conclusions from our theoretical structure than has been possible in the past. Eddington has made valuable contributions along these lines. But surely it is a distortion of the case to conclude from these quantitative deductions that the deductions are a result of "the primitive forms of thought working themselves out." (PPS, p. 178) He is much nearer the truth when he writes:

> We have to show, not that there are N particles in the universe, but that anyone who accepts certain elementary principles of measurement must, if he is consistent, think there are. A logically complete demonstration, if it is possible, would be extremely prolix; and it is not the kind of problem I could myself attempt. But I shall try to show that at each stage the investigation is being driven by its own momentum – that the moves leading to a universe of N particles are forced. Or at least there is so much pressure behind the moves that, when we find the physicist actually does think there are N particles, there can be no doubt that it is the result of this pressure and not because of any peculiarity in the external world.[1]

Can this kind of push from the observer's point of view yield the kind of necessity and *a priori* prediction Eddington claimed? Does not the force of his pretended Kantian hypothetical deduction depend upon a "logically complete demonstration," which is what he precisely declines to give? The answer is that, if he is

---

[1] "The Evaluation of the Cosmical Number," *Proc. Camb. Phil. Soc.*, vol 40, 37; Reprinted in *Fundamental Theory*. "Eddington proceeds much more by analogy than logic." (p. 4, Slater, *op. cit.*)

to make good his claims for an *a priori* of method, he is committed to producing just those complete *a priori* deductions which he never does produce. Slater testifies that, in the early days of his work on his *Fundamental Theory*, "Eddington amended his calculations, where they did not fit closely to observation, by refining and complicating his (mainly dimensional) arguments until they did fit."[1] He has tried to twist the subjectivist features of the quantum world into an epistemological doctrine of *a priori* knowledge. The world of physics is a complex world from which the observer can no longer be abstracted. But these subjectivist features do not warrant so extravagant a claim as that advanced by Eddington. Neither his claims concerning the significance of the number 4, nor the peculiar role played by the observer in quantum physics, leads to the *a priori* doctrine advanced by Eddington. He may have developed a new mathematical deduction, this is for competent mathematicians to judge; but he has not discovered any *a priori* path to knowledge.

Many other contemporary physicists have reflected upon the philosophical changes entailed in physics by relativity theory and by the new advances in quantum mechanics. There seems to be general agreement that the old world view of an absolute reality existing independent of analysis and discovered through analysis, must be replaced by the phenomenalist world of the observer and his observations. The acceptance of the operational point of view has illustrated this change. Some scientists, such as Eddington, have studied the inseparable role of the observer in the results of science and have concluded that the resulting knowledge should no longer be called objective, without explicit qualifications. Jean Mariani developed these views in France at about the same time as Eddington was making them popular in England.[2] Mariani gave Eddington credit for being one of the the first to introduce the notion of subjectivity as a means of accounting for natural laws. But his own revision of the traditional views is not so radical as Eddington's. His main contention is that in quantum theory we find "*a theory in which the objective laws of nature do not translate the objective properties of objects.*" (p. 26) He stresses the distinction between macro- and microscopic

---

[1] *Op. cit.*, p. 4.
[2] *Les Limites des Notions d'Objet et d'Objectivité*, 1937.

levels but tries to maintain his subjectivist contention even on the level of social science, in psychology and sociology. Objectivity becomes for Mariani the invariant factor, dependent upon the choice of the observer for a particular reference system.[1] J. L. Destouches has followed Mariani's lead but he has carefully dissociated himself from the views of Eddington. While accepting the subjectivist revision of traditional physics, he is not advocating the radical position of thinking to find "in the laws of that Physics only our own mental structure."[2] Like Eddington, Destouches has been concerned to set up a deductive system, but he has been more precise than Eddington about the structure of such a system. Basic axioms providing the grounds for deductive steps throughout the system constitute one of the more fundamental features of any deductive formulation.

> The end followed by the deductive method is to replace discussions in each particular instance by general rules announced at the beginning of the theory, rules of definition and justification. These rules enable us to introduce the maximum of connections so that the schema will be adequate: knowledge of the beginning of the connections enables us then to know the remaining connections by using the rules.[3]

Like Mariani, Destouches makes the results of physics apply to the phenomenalist situation, his deductive system being a formulation of the observational knowledge of the scientist. The results of measurements cannot be properties intrinsic to objects. "A physics in which all the simultaneous measurements are not possible is not able to be a physics of intrinsic properties but must limit itself to being a physics of relations."[4] Within this relational system, predictions can be made but only by starting from precisely described measures. The whole system of predictions is based, in other words, upon the elements taken as members of groups and upon their relations in the system. Destouches's deductive system pertains, that is, to those contexts which can satisfy the mathematical criterion of structure; the theory of groups forms the system within which he finds it possible to make predictions from basic axioms of the system. It is with this

---

[1] Cassirer has also identified objectivity with invariance. His special brand of critical idealism is developed in his *Substance and Function*.
[2] *Principes Fondamentaux de Physique Théorique*, t. 1, p. 9.
[3] *Essai sur la Forme Générale des Théories Physiques*, p. 2.
[4] *Ibid.*, p. 102.

general theory in mind, that Mme. Destouches-Février has sought
to extend the concept of predictions to more philosophical situations. Assuming the goal of scientific theory to be "to predict,
starting from the initial results of measurements, the results of
future measurements," she outlines the essentials of her theory
of predictions.

> Starting from this idea, we are able to construct a formalism called
> a *general theory of predictions* which, in virtue of the introduction of a
> certain number of concepts and principles, enables us to deduce a great
> part of the properties of physical systems.[1]

The theory of predictions holds good only for the microscopic
level where the tools of measurement are inseparable from the
results of measurement. The predictions within the deductive
framework are dependent upon the actual situation in this field
of physics. The deductive structure does not, by itself, introduce
the element of prediction.

The observer does play a unique and indispensable role in obtaining knowledge on the microscopic level of quantum physics.
The introduction of a deductive formalism into the results of
quantum physics does make possible predictions from one part of
the formalization to another. Eddington's error was his failure to
recognize that such predictions could be made only within a
well-defined and carefully constructed context. This context is
the result of a definite conceptual framework used in dealing with
quantum phenomena; the concepts of this framework control the
explanatory theories on this level.[2] Once we admit the validity
of predictive systems of the sort discussed by Destouches and
Mme. Destouches-Février, we can give to Eddington's claim
more justification than is usually accredited to it; for the *a priori*
claim made by Eddington for both the macro- and microscopic
levels reduces to something very like the deductive systems of
these writers. Mme. Destouches-Février's distinction between
these two levels in terms of the difference in the rôle of the observer is too sharp if taken as a difference in kind; for the difference is only one of degree: the observer does appear in the results
of measurement and observation even on the gross level. In fact,

---

[1] "Monde Sensible et Monde Atomique," *Theoria*, t. XV, pt. I–III, 1949, p. 81.
See also *Dialectica*, t. 1, 1.
[2] Vide N. R. Hanson, *Patterns of Discovery*, Cambridge, 1959.

Milne has constructed a deductive system based just on the fact of the integral part the observer plays in modern physical knowledge, a construction which finds its correlates in Destouches's *Principes Fondamentaux de Physique*. Whitrow has drawn in the contrast between the old and the new physics which makes Milne's and Eddington's rationalistic stress possible.

In the past these 'observers' have been regarded as mere spectators whose role was to act as judges in the final appeal; but today, both in relativistic and in quantum physics, natural science is regarded as an activity rather than as an object for Platonic contemplation. Consequently, observers tend to become witnesses who themselves directly assist in determining the nature of the eveidence.[1]

The alteration in the meaning of 'invariant,' from that which is timeless to that which is the same for all observers of a specified class, is evidence of this changed attitude towards the role of the observer even in macroscopic physics. Milne's Kinematic Relativity is the most radical embodiment of this change. His relativity theory is based, Whitrow points out,

on the *abstract* concept of an observer, or mental monad, who experiences a temporal before-and-after sequence of events. The essential feature *defining* this sequence is its *irreversibility*. The co-existence of one other similar observer and of a signalling process of intercommunication is sufficient to give initial content to the time concept, provided the two observers do not coincide at all epochs.[2]

With this point of departure, Milne and his followers believe they can construct a rigid deductive system containing all the laws of nature commonly recognized by scientists. But although Milne's system is generally admitted as being more sound than that of Eddington, it has received much the same sort of negative criticism. The reason for this is again the term 'a priori,' for Whitrow and Milne insist that they have discovered an *a priori* system.

If our concept of scientific method implies congruent measurements by a continuum of hypothetical observers, the *initial* choice of a particular geometry as an ideal background, against which physical phenomena are to be silhouetted, can be decided by *a priori* epistemological considerattions.[3]

[1] "The Epistemological Foundations of Nature Philosophy," in *Philosophy*, vol. XXI, 1946 p. 21.
[2] *Ibid.*, p. 22.
[3] *Ibid.*, p. 11.

What they have been able to do is to deduce the normal laws of nature and other principles of physics from a few simple premises. They have, in other words, been able to establish a mathematical deductive system for physics, starting from premises different from those invoked by Eddington, or, for the most part, those invoked by Einstein. Kinematical equivalence becomes the most basic concept.

> Two particle observers, A, B are said to be *kinematically equivalent...* when the totality of observations that A can make on B can be described by A in the same way as the totality of observations that B can make on A can be described by B. (A and B are supposed to be provided with clocks and to be capable of sending light-signals to one another).[1]

Milne seeks to show that the laws of dynamics can be derived from this and a few other factors, stressing the purely rational character of the endeavor.

> No appeal is made, in the deduction, to any specific theory of gravitation, or to the supposed existence of any universal constant of gravitation. The inference is that the phenomenon of gravitation does not depend on the micro- or macro-structural properties of matter, but is an inevitable element in the motion which particles undergo in one another's presence if they are to be consistently observed by observers in a universe satisfying Mach's principle.[2]

In short, he maintains that he has been able to derive the laws of dynamics and the Newtonian approximation to the law of gravitation from a kinematic basis.

> That is, they have been deduced rationally, starting from the individual observer's awareness of a temporal sequence for events at himself, and from his assigning of measures of distance and epoch by means of light-signals and appropriately rated and synchronized clocks; explicit procedure for this rating synchronization having been stated in terms of the observer's own experiences.[3]

The actual laws which phenomena obey can be inferred, Milne believed, from the definitions of the system he has laid down. His followers more than himself have insisted upon calling this system *a priori*; but he stresses the rational nature of kinematic relativity.

[1] Milne, E. A. "The Inverse Square Law of Gravitation," *Proc. Royal Soc.*, A156, p. 66.

[2] *Ibid.*, p. 81. Mach's principle is that the inertia of any particle of matter has its final explanation in its reference to all the other particles of the universe.

[3] Milne, E. A. "Kinematics, Dynamics, the Scale of Time," *Proc. Royal Soc.*, A156, p. 324.

We begin with local phenomena, which are shown to be possible and self-consistent, and we can end with local laws, and whether the actual universe follows the details of the extrapolation is quite immaterial. We simply require that local phenomena and local laws be such that they are capable of being fitted together to form a complete universe.[1]

The task of observation is to identify the entities in nature as corresponding to those factors of the theory. The laws of nature are thus axioms in a coherent system, and "observation vindicates not the truth of the theorems (which depend for their validity only on the logical self-consistency of the original axioms and the absence of errors from the reasoning connecting the theorems with the axioms) but the applicability of the axiomatic definitions to the things occuring in nature."[2]

Milne's system has not received any more general acceptance than has Eddington's abstruse claim for the deduction of the constants of nature. Both systems have been distrusted because of the emphasis by their authors on the *a priori*, rationalistic aspect of their systems. Proof of the integral rôle played by observers in modern physics does not wait for the vindication of Milne's kinematic relativity, for specification of reference frames or the location or perspective of observers is a fundamental necessity in relativity theory. Construction of deductive systems of our present physical knowledge can be done from varied initial axioms. If Milne has succeeded, as he and his followers firmly believe, in constructing such a system based upon the postulate of two observers exchanging light signals, we need not shy off because we feel such a claim violates the experimental dictum of science. As Schrödinger admitted in reviewing Eddington's *The Philosophy of Physical Science*, laws of physics frequently pass from empirical generalizations to tautological assertions, although he hangs back from the view, which he credits to Eddington, that

the content of the theory [of realitivity] has been switched over ... from empirical regularities to essential ingredients in our method of attack on problems of physical science and had thereby been both deprived of the benefit and withdrawn from the danger of being checked by experiment.[3]

[1] *Ibid.*, p. 326.
[2] *Ibid.*
[3] *Nature*, vol. 145, March 16, 1940, p. 402.

It was the apparent indifference of both Eddington and Milne to the empirical referents of their systems which aroused much of the antipathy against them. Reviewing Eddington's *The Philosophy of Physical Science*, Dingle insisted upon every system's (whether deductive or not) being directly related to experience.

Now with regard to the *a priori* derivation of the laws, it is possible here only to say that if a rational system is not essentially a correlation of *experiences*, it has no scientific interest. It is well known that an infinite number of rational systems may be constructed if the premises are uncontrolled, and the chance that a pure *a priori* system will happen to correlate experience is negligible. If Sir Arthur has indeed found a system which does, we must suspect the premises – 'his frame of thought' – and, sure enough, we find experience there.[1]

Judging from this remark, Dingle understands by the term 'a priori' something completely independent of experience, something spun by the theorist with no basis in experience. He pictures Eddington as claiming that by sitting in his armchair, with no reference to existing empirical data, he has been able to construct a deductive system which takes account of all the important laws of nature. Dingle believes that the fact that a close analysis of Eddington's book reveals a vast substructure of experience indicates that Eddington has not really constructed an *a priori* system. Unfortunately, this concept of *a priori* systems seems to have been shared by a number of other scientists who accordingly followed Dingle's lead in rejecting Eddington's claims. But if Dingle had paid close attention to all aspects of *The Philosophy of Physical Science*, he would have seen that Eddington does not mean by *a priori* any such transcendental system as a pure intelligence might create. True, he claims that the deductions he has made could have been foreseen by a superior mind totally unacquainted with the empirical findings of modern science, but statements of this sort are obvious elliptical remarks. Eddington defines his use of 'a priori' quite early in this book. "I think that I am using the term *'a priori* knowledge' with its recognized meaning – knowledge which we have of the physical universe prior to actual observation of it." (PPS, p. 24) He goes on to specify that "*a priori* knowledge is prior to the carrying out of observations, but not prior to the development of a plan of observation. As physical knowledge, it is necessarily

[1] *The Observatory*, vol. 63, 1940, p. 22.

an assertion of the results of observations imagined to be carried out." (*Ibid.*) And he concludes that *a priori* knowledge can not be said to be independent of experience. "We must grant then that the deduction of a law of nature from epistemological considerations implies antecedent observational experience." (PPS, p. 25)[1]

Eddington's *a priori*, then, does not operate in isolation from experience or from the empirical data of science. The empirical data are requisite for the epistemological deductions: the data provide some of the basic axioms for the deductive formulation.

He infers the numerical constants from the broad principles of physics; these principles he would like to ascribe to 'the system of thought by which the human mind interprets to itself the content of its sensory experience' (*Relativity Theory of Protons and Electrons*, p. 327). But the sensory experience is a vital part of the knowledge; and even if the physical principles assume a qualitative appearance when formulated... their background is in quantitative measurement.[2]

The pure *a priori* criticized by Dingle cannot be found in Eddington's analysis. Schrödinger[3] called attention to an important remark made by Eddington about his *a priori* claim: "I have been acting as an advocate for an extreme view, presuming that your natural prejudices are all the other way." (PPS, p. 113) Schrödinger suggests that in the light of this remark, we should not interpret Eddington in too literal a manner. But even setting this remark aside, it should have been clear to readers of *The Philosophy of Physical Science* that Dingle's criticism were invalid, that the *a priori* system promised by Eddington was nothing more than a deductive formalization of science.

The same conclusion became evident also from the perceptive criticisms of Sir James Jeans. Jeans distinguished between two different sorts of epistemological premises, those which are epistemological in form only – appear to be about methodology in abstraction from data – and those which are in substance epistemological in Eddington's sense. Jeans argued that Eddington had not produced premises which were epistemological in substance. Of the following three propositions claimed by Eddington

---

[1] Cf. H. Jeffreys, *Theory of Probability*, 2nd ed. 1948: "Eddington's ¦starting point is not purely epistemological because it assumes the considerable amount of observational evidence that is needed before we can establish a measuring system at all." (Quoted by Slater, *op. cit.*, p. 288.)

[2] Slater, p. 3.

[3] Review of PPS, in *Nature*, vol. 145, pp. 402–03.

as being *a priori* (in Jeans's terminology, epistemological in substance), Jeans argued that only the third is properly epistemological and that only in form. (a) The laws of nature involve only relative velocities. (b) It is impossible to determine an absolute velocity in space. (c) It is meaningless to talk of simultaneity at distant points. Eddington claimed that proposition (c) could have been foreseen even if the Michelson-Morley experiment had not taken place, thereby making this proposition epistemological in substance, that is, *a priori*.

> The Michelson-Morley experiment did more than instigate a scrutiny; it disclosed one of the 'brute facts' of Nature, and it is on this fact that proposition (c) depends for its very existence; if the experiment had not turned out as it did, there would have been no proposition (c). Thus (c) is not *a priori* knowledge in the sense of the rationalist philosophers, or even in the sense of Eddington ...[1]

Jeans claims that the same argument holds good for all the other supposedly epistemological propositions cited by Eddington: they can all be stated so that they have the form of a priority but never so that they are substantially so. Eddington's reply to these charges is helpful in showing that he was not advancing a doctrine of a pure *a priori* but was only calling attention to what he believed were genuine deductive possibilities within the empirical sciences. He charged Jeans with being outmoded in his statement of the Michelson-Morley experiment.

> Let us try to express this 'brute fact' not in the terminology of 1887, or even of 1905, but as it appears in 1941. To say that the experiment will give a null result if strains are properly eliminated has become a tautology. The new fact is that, with the experimental precautions adopted, strains *are* eliminated.[2]

Eddington did not claim that the deductions he thought could be made were possible prior to the twentieth century, or even prior to events of 1905. The situation in contemporary physics has produced a tautology where formerly there were only empirical data. But this does not mean that the Michelson-Morley experiment was useless or that its only use was to speed up the recognition of the contradictions entailed in the attempted measurement. What Eddington should have said was that now,

[1] Jeans, *Nature*, vol. 148, p. 140.
[2] *Ibid.*, p. 141.

after the experiment and after the development of relativity theory with its distinctions of reference frames and coordinate systems, a deduction can be established which can ignore the experiment in its formulation. Eddington's careful placing of his claim in 1941 is significant even though he does not express the full meaning of this fact. He drew a distinction between pointing and describing which he thought helped to clarify the difference between Jeans and himself, but this only goes part way towards a proper clarification. After Jeans had replied to Eddington's answer, repeating the same charge, Eddington restated his distinction.

> It is a logical impossibility that the Michelson-Morley experiment should give a null result in the conditions *described*; but the possibility imagined is that a bogey, supposed to have been laid, has come to life again, so that the conditions described are not those which have been *pointed out* to the experimenter.[1]

Attempts to describe the conditions under which the measurements would be conducted end, when all the known data and theory are taken into account, in the recognition that it is impossible.

No *a priori* deduction has been made: there has only occurred a formulation of the empirical results in a more rational, logical order. Jeans was certainly correct in urging the importance of the Michelson-Morley experiment and insisting that the deduction or conclusion stated by Eddington could not have been made independent of this empirical experiment. It could not have been made prior to the experiment, although it can now be stated without explicit reference to that experiment. Eddington's deductive theory constantly assumes, what Eddington himself sometimes forgot, all of the empirical data of modern science. His program was not, as Dingle suggested in his letter in *Nature* with reference to the Jeans-Eddington dispute, to construct a scheme of pure reason based on a few *a priori* postulates in terms of which observational data are described.[2] The postulates of his system were simply the empirically discovered and verified principles of relativity theory from which he proceeded to show how a deduction of the rest of physics could be made. To the extent that much

[1] *Ibid.*, p. 256.
[2] *Ibid.*, p. 341.

of relativity theory has undergone only a very indirect check by experimentation, and to the extent that relativity physics rests upon the theoretical framework of Einstein and others, we might be able to say that modern physics is 'pure' in Dingle's sense; but it was not this aspect of the accepted new forms of physics which bothered Dingle. Eddington did not help his position by arguing, in his reply to Dingle's letter, that the laws of physics do not have any special application to experience such that a violation of the law would indicate a necessary error in the law. He was correct in pointing out that violations of laws are usually interpreted as psuedo-violations and some errors in calculation, observation, or object thought to be observed are suspected. But he argued his case in terms of a coherence theory.

I think then that the only necessary condition is that the fundamental laws shall form a single scheme, applicable to experience in the sense that all varieries of knowledge fathered by the methods of physical science can be formulated in terms of it. The choice is not limited by considerations of simplicity of application, utility, or appropriateness to the actual state of the universe, and is therefore left open to be determined by the *a priori* considerations described in my last letter.[1]

Dingle insists upon the reference to experience as being fundamental, and yet in his own work, *Through Science to Philosophy*, Dingle argued that reason places interpretations upon experience and that the choice of a postulate of interpretation is made in terms of its coherence with the other postulates in the set.

Interpretation is the pineal gland of the reason-experience duality. It obeys no laws – there is no reason why any postulate should not be adopted as a symbol of any experience. It appears to be as essential to the nature of consciousness as is time: memory petrifies experience and makes it amenable to rational treatment, and interpretation converts the petrified memory into a definable concept. (p. 103)

The truth of any memory experience is tested *"by seeing if it forms a rational correlation with other memories."* (p. 186).

Dingle strangely failed to recognize in the systems of Milne and Eddington the same doctrine of interpretation of experience as he himself advocated. Misled by Eddington's extravagant statements, he saw both men advancing a doctrine of pure *a priori* knowledge. Not so sympathetic as Eddington and Milne to mathematics and its use in modern science, he failed to ap-

[1] *Ibid.*, p. 342.

preciate the unique contributions of this science to the physical sciences. In their deductive systems he saw a challenge to the very foundations of science: are the foundations to be observation or invention? "Instead of the induction of principles from phenomena we are given a pseudo-science of invertebrate cosmythology, and invited to commit suicide to avoid the need of of dying."[1] But Dingle was incapable of reading Eddington sympathetically, *The Philosophy of Physical Science* being particularly difficult for him.[2] But beneath the many misleading statements made by Eddington in the presentation of his doctrine, there lies the essentially valid insight into the nature of explanation in the sciences. It is true, as G. Burniston Brown pointed out, that Eddington was concerned chiefly with the exact mathematical sciences and not with physical science in general.

His suggested method is to start with a symbolic structure of great generality and build up other symbols having various relationships. He then expects the scheme of relationship arrived at by the analysis of *metrical* observational knowledge to be identifiable among the plethora of schemes arrived at *a priori*.[3]

But although Eddington's system finds its origins in the exact sciences where pointer readings are the sole data and mathematics the basis for all formulations, he clearly sought to extend his doctrine to cover the whole of physical knowledge. He was concerned with the mathematicization of science. Dingle raised the general question of the relation of mathematics and experimental physics, of rationalism and experience.

Norman Campbell expressed general agreement with Eddington's and Milne's contention that explanation is basically deductive. The process, Campbell says, has three stages.

The formulation of hypotheses involving the acceptable ideas alone, the deduction of consequences from these hypotheses, and the translation of these conclusions into propositions concerning the less acceptable ideas

[1] "Modern Aristotelianism," in *Nature*, vol. 139, p. 786.

[2] Dingle's summation of this book is a good illustration of the emotionalism aroused in him by Eddington's *a priori* claims. "To sum up: this amazing book, like Aristotelian tragedy, purges the emotions through pity and terror – pity that such supreme scientific achievements should be disguised in so impenetrable a mask of untenable philosophy, and terror at the prospect of having to see through the mask or miss the science." (*The Observatory*, vol. 63, 1940, p. 25.)

[3] *Nature*, vol. 148, p. 504.

by means of a 'dictionary.' The peculiarity of scientific explanations is that they often (not always) predict new laws in addition to explaining old ones.[1]

The fact that in the cases of Milne and Eddington we have explanations in purely mathematical terms has resulted, Campbell argues, simply from the fact that these men have concerned themselves with laws which have no mechanical model and are very abstract. Milne and Eddington were not, however, so much concerned with predicting new laws as with accounting for the old ones within a deductive system. Milne, in replying to Dingle's charges of Aristotelianism, commented that his position evolved from asking what are the consequences of the assumption "that the universe is on the average homogeneous."[2] He insisted that such an assumption "is not an *a priori* belief to be scoffed at; it is a fact of experience to be reckoned with, that when we do thus eliminate such empirical appeals, regularities emerge (as logical consequences of the hypothesis) which play the part of the very laws of Nature which are *observed* to hold good."[3] He recognized clearly that the "trend of physical investigation is at once to extend the number of known facts by empirical discovery, and to diminish the number of independent facts by establishing some as consequences of others."[4] The development of the second trend leads to turning physics into geometry where the laws of nature become theorems. "The role of observation would again be to attempt to verify in Nature the existence of entities corresponding to those mentioned in the axioms ... by ascertaining whether the resulting theorems hold good in the external world."[5] Milne was aware that the results of such deductions may not find any correlations in experience, implying that if such is the case he would willingly abandon the deductions, alter his postulates, and make new calculations. He believed that the discovery of laws through induction leaves the laws irrational, for they cannot be explained. If they can be deduced within a logical system we can clothe them with rationality. The alternative to inductively derived laws is that the universe is rational,

[1] *Nature*, vol. 139, p. 1005.
[2] *Ibid*, p. 999.
[3] *Ibid.*, p. 998.
[4] *Ibid.*
[5] *Ibid.*

can be formulated in a coherent system of deductions. Reviewing Bridgman's *The Nature of Physical Theory*, Milne lays the background for his later judicious reply to Dingle. Arguing that "the revulsion from the Aristotelian or *a priori* method to the Baconian or inductive method has led to the under-estimation of the importance in present-day science of the discovery of new thought processes," Milne made quite clear that the development of modern science is from inductive to deductive procedures.[1] Knowledge of the external world is had via observation but understanding of that world cannot be acquired in the same way. Understanding arises only when we have been able to arrange the data of observation into a coherent, deductive system with a minimum of postulates. Like Kant, Milne was concerned with explaining experience by ordering its ingredients into a logical system. More forcefully than Eddington, Milne has expressed the motivating idea behind both of their attempts at what they have poorly labeled an *a priori* system. The empirical nature of science has not been violated: the theoretical, rational side has merely been placed in a prominent light.

---

[1] *Math. Gaz.*, vol. 20, 1936, p. 341. Cf. M. Polanyi's "From Copernicus to Einstein," *Encounter*, Sept. 1955, pp. 54–66, where he insists upon the rôle of rationality and intelligibility as important guides in the foundation of scientific theories.

# CHAPTER VI

# THE CONCEPT OF REALITY

In our survey of Eddington's philosophical interpretation of the physical sciences, we have found two fundamental strands at work, the operational-phenomenalist trend and the causal theory's realist assumptions. Men like Dingle have argued that these two points of view are antithetical, that they cannot be brought together, that, in fact, the presence of the latter indicates that Eddington just did not succeed in throwing off all vestiges of the older absolute position. The "inscrutable 'conditions of the world'" of his *The Mathematical Theory of Relativity* hung, Dingle protested, "like the Old Man of the Sea round the neck of his thought, contributing nothing and serving only to retard its progress and obscure results which, expressed simply and directly in terms of the essential measurements alone, might have commanded understanding and acceptance."[1] What Dingle considered a constriction to the full development of Eddington's thought was, however, a conscious part of Eddington's system. The causal theory of perception and the doctrine of structure were integral aspects of his interpretation of the sciences introduced to account for specific problems of physics and epistemology which the operationalist refuses to consider. It can be argued that the operationalist-phenomenalist language is adequate to express the actual procedures of science, although those sympathetic to this point of view have of late been arguing that, even as a linguistic formulation of the processes of science, such a language is defective and insufficient. I have argued in chapter I that, even though we accept the operationalist position as adequate for expressing the findings of science, a vast non-operational context is presupposed by science. Eddington was aware, though not always adequately, of this necessary context within which the sciences have their meaning and significance. His own philosophy of science was concerned with laying bare

[1] *Proc. Phys. Soc.*, vol. 57, p. 246.

the epistemic presuppositions of the methods and procedures of the physical sciences. Such a task took him into the causal theory and led to his elaboration of the doctrine of structure in its perceptual form, as well as to the development of his purported *a priori* claims. Analysis of the epistemological presuppositions of science, revealed, in other words, certain crucial ontological assumptions indissolubly associated with the epistemological background. There is a rising emphasis in Eddington's works on the subjective, *a priori* claim, an emphasis which almost usurps all other problems in *The Philosophy of Physical Science*; but even in that work the doctrine of structure plays an important part in the exposition of his position. Along with the *a priori* claims and the general theory of relativity go the ontological presupposition of a subjective phenomenalist reality, a reality shaped and determined by the categories and methodology used by the observer. For the doctrine of structure and the causal theory there is presupposed a realist concept of reality. Both realist and phenomenalist ontologies are present throughout Eddington's writings. Even though I have argued that the *a priori* claims are misplaced and reduce to the valid claim of organizing science into a deductive system, the phenomenalist ontology is still an important facet of Eddington's general interpretation of the sciences.

Just as the operationalist statements contain objections to the older Newtonian non-operational position, so the elaboration of the subjective concept of reality affords frequent opportunity for fulminations against the 'metaphysical' views of older ontologies.

By defining the physical universe and the physical objects which constitute it as the theme of a specified body of knowledge,... we free the foundations of physics from suspicion of metaphysical contamination. This type of definition is characteristic of the epistemological approach, which takes knowledge as the starting point rather than an existent entity of which we have somehow to obtain knowledge. (PPS, p. 3)

His major emphasis in these Tarner lectures was upon the subjective concept of reality, since this concept is required by his *a priori* claim. He was well aware that the *a priori* claim could not have a chance of success unless he could show that the world of physics was the world determined by the tools (sensory and intellectual) of the observer. We have seen that this subjectivist position is not dependent upon the *a priori* claim. In fact, it was undoubtedly the subjectivist features of relativity theory, especially as they occur in quantum physics, which led Eddington to formulate the epistemic presuppositions of science as *a priori*. The result of his reflections upon the relativistic features of modern physics led him to the conclusion that we are "unable to reach by physical methods a purely objective world, and it would seem to follow that all the entities of physics have the partial subjectivity of the world to which they belong." (NPS, p. 292) When taken out of the context of the *a priori* claim, the subjectivist concept of reality assumes a distinctively phenomenalist character. Even in *The Philosophy of Physical Science* we find the distinction between particular or individual subjectivity and generic subjectivity, the latter being dependent upon the sensory and intellectual equipment of the observer, while the former is dependent upon his position, velocity, acceleration, and other variable factors. Objectivity is used in the sense of that which is universal, common to all observers. (PPS, p. 87) *New Pathways in Science* opens with a quotation from Poincaré's *The Value of Science* to the same effect. "What we call 'objective reality' is, strictly speaking, that which is common to several thinking beings and might be common to all; this common part, we shall see, can only be the harmony expressed by mathematical laws." One of the uses of the term 'absolute' is that of universality, as erg-seconds are said to belong to "Minkowski's world which is common to all observers." (NPW, p. 180) 'Absolute' in the sense of 'universal' with respect to all observers is the usual meaning of the term in Eddington's books, although he occasionally speaks of the absolute scheme of things as if this absoluteness were independent of observers. (NPW, p. 38) In *Space, Time and Gravitation*, he defines 'absolute' as "a relative which is always the same no matter what it is relative to. Although we think of it as self-existing, we cannot give it a place in our knowledge without setting up some dummy to relate it to." (p. 82) But universality may mean either that common area of agreement in perspectives or the independent world known by the observer. Both shades of meaning are found in Eddington's dual concept of reality. The most frequent form taken by the phenomenalist use of 'absolute' (that common to all perspectives) is the constructivist formulation used by Russell in his *Analysis of Matter*.[1] Here "the external world of physics is ... the symposium of the worlds presented to different view points." (NPW, p. 284) Like Russell, Eddington recognized that "the only subject presented to me for study is the content of my consciousness," and that the concept of reality must be developed from this core of indubitable data.

Accordingly my subject of study becomes differentiated into the contents of many consciousnesses, each content constituting a *view-point*. There then arises the problem of combining the view-points, and it is through this that the external world of physics arises. Much that is in any one consciousness is individual, much is apparently alterable by volition; but there is a stable element which is common to other consciousnesses. That common element we desire to study, to describe as fully and accurately as possible, and to discover the laws by which it combines now with one view-point, now with another. This common element cannot be placed in one man's consciousness rather than in another's; it must be in neutral ground – an external world. (NPW, pp. 283–84)

The hypothesis of externality becomes useful only when "it is the means of bringing together the worlds of many consciousnesses occupying different view points." (*Ibid.*, p. 284) The synthesis of all possible points of view, together with the rules of the synthesis, constitutes the scientific concept of reality in this phase of Eddington's thought.

When we (scientists) assert of anything in the external world that it is real and that it exists, we are expressing our belief that the rules of the symposium have been correctly applied – that it is not a false concept introduced by an error in the process of synthesis, or a hallucination belonging to only one individual consciousness, or an incomplete representation which embraces certain view-points but conflicts with others. (NPW, p. 285)

But he admits that it is consistent with this constructivist interpretation to say that the resulting construct is not the only possible world really existing. (NPS, p. 26) He suggests both that other perspectives than the human might yield a different constructed reality and that a world independent of all perspectives

---

[1] Dobbs has expressed the constructivist concept recently. "In short the idea of a 'public' world of physical events is a construction based on the covariance found in the experiential events of differently situated and constituted observers, discovered by the inter-observer communication achieved through the medium of language." ("The Relation Between the Time of Psychology and the Time of Physics," BJPS, vol. 2, 1951, p. 128.)

might exist (his 'conditions of the world'). But in *Space, Time and Gravitation*, the constructivist concept is made to apply to a very wide area. He argues that no valid reasons can be given for taking any one appearance or perspective as more real than any other: "reality is only obtained when all conceivable points of view have been combined." (p. 182) The synthesis here is made to include extra-terrestrial observers. The real becomes defined as the product of a synthetic merging of all points of view in order to sift out the particular and personal. Observability determines the scope of the application of the concept of reality. Operationalism and phenomenalism fit neatly into the constructivist concept, for the synthesis of all possible points of view follows the tradition laid down by Berkeley and Mill, the latter of whom suggested that matter itself is a permanent possibility of sensation. The common area shared by all individuals in Eddington's constructivist concept fulfills many of the functions of Price's family of sense-data. As Price states the theory of phenomenalism, it is the theory which posits as particulars only sense-data, but it does not assert that a sense-datum is a material thing.

What it says is that a *system* of sense-data is a material thing; and not any system you like, but that very special sort of system which we have called a family – a system, too, which contains obtainable sense-data along with actual ones. The causal characteristics necessary to matter, e.g., impenetrability, belong according to it not to any single sense-datum, but only to the whole family collectively.[1]

Eddington did not work out the details of his constructivist concept; but had he done so he might have arrived at a formulation similar to Price's account of phenomenalism. The phenomenalist concept of reality remains in favor among those philosophers of science who lean towards positivism and operationalism. One half of Eddington's thought belongs to this tradition.

No matter how extensive we wish to make such a concept of reality, Eddington has insisted that the constructivist concept presupposes a non-constructed reality. Operationalism takes on meaning only within the non-operational context which surrounds the operations of science. "Selection implies something to select from. It seems permissible to conclude that the material on which the selection is performed is objective." (PPS, p. 26) The ultimate

[1] *Perception*, pp. 282–83.

operand "must be free from subjectivity." In many places he claims that

The object of the relativity theory ... is not to attempt the hopeless task of apportioning responsibility between the observer and the external world, but to emphasize that in our ordinary description and in our scientific description of natural phenomena the two factors are indissolubly united. (STG, p. 33)

Although he insisted that the task of determining the kind of reality existing independent of all observers is outside the scope of physics (STG, p. 31), he did not shun the problem in his own interpretation of the sciences. He was convinced that direct knowledge of such an objective world was for the most part impossible, but he formulated his doctrine of structure as a way of sifting the subjective from the objective ingredients of scientific knowledge. We have seen his strong insistence in *Mathematical Theory of Relativity* upon the formulation of science being related through structure to a world independent of all observers. Early in *Space, Time and Gravitation* he argues that "Although length and duration have no exact counterparts in the external world, it is clear that there is a certain ordering of things and events outside us which we must find more appropriate terms to describe" (p. 35) When he does seek to describe the counterpart, he tries to assimilate it to his constructivist concept.

Although different observers separate the four orders differently, they all agree that the order of events is four-fold; and it appears that this undivided four-fold order is the same for all observers. We therefore believe that it is inherent in the external world; it is in fact the synthesis, which we have been seeking, of the appearances seen by observers having all sorts of positions and all sorts of (uniform) motions. It is therefore to be regarded as a conception of the real world not relative to any particular circumstanced observer. (STG, p. 36)

In the same manner, force "depends upon the acceleration of the observer holding the balance" which measures the force, and, like length and duration, may have no counterpart in the external world. He adds, however, that "There is, of course, something at the far end of the link, just as we found an extension in four dimensions at the far end of the relations denoted by length and duration." (*Ibid.*, p. 43) The choice of the phrase 'at the far end of the relations' is hardly consistent with the constructivist concept of reality but suggests again that he was trying to make his realist views harmonize with his phenomenalist concept. Interval is another factor said to possess objectivity because it posits no particular observer. (*Ibid.*, p. 46) Likewise, the speed of light is absolute in the sense of being the same for all observers. (*Ibid.*, p. 59) Geodesics are said to be absolute for similar reasons, when applied to the path of freely falling particles.

> Our attention is thus directed to the natural tracks of unconstrained bodies, which appear to be marked out in some absolute way in the four-dimensional world. There is no question of an observer here; the body takes the same course in the world whoever is watching it. Different observers will describe the track as straight, parabolical, or sinuous, but it is the same absolute locus. (*Ibid.*, p. 70)

The track of freely falling particles can be measured absolutely since all observers will agree concerning this measurement.[1] It is never completely clear in these passages whether Eddington looked upon these factors as absolute and objective because they were the result of the synthesis of points of view (his constructivist concept) or because, being the same for all observers, they reveal a world independent of observers. Later in the same book, he argues that, although gravitational force is relative to each frame of reference, there is a complex character of gravitational influence which is absolute in the sense of being invariant for all frames of reference. He here speaks of "the presence of a heavy particle" as modifying "the world around it in an absolute way which cannot be imitated artificially," strongly suggesting that we are dealing with that ultimate operand which he speaks of in his later work. The separation of the constructivist and realist concepts of reality becomes even clearer when he speaks of the intersection of world lines.

> There is one type of observation which, we can scarcely doubt, must be independent of any possible circumstances of the observer, namely a complete coincidence in space and time. The track of a particle through four-dimensional space-time is called its world-line. Now, the world-lines of two particles either intersect or they do not intersect; the standpoint of the observer is not involved. In so far as our knowledge of nature is a knowledge of intersections of world-lines, it is absolute knowledge independent of the observer. (STG, p. 87)

[1] In NPW, the role of geodesics is clearly assigned to the realist concept. "There are certain curves which can be defined on a curved surface without reference to any frame or system of partitions, viz., the geodesics or shortest routes from one point to another. The geodesics of our curved space-time supply the natural tracks which particles pursue if they are undisturbed." (pp. 124–25)

He ends the same book by suggesting that the law of atomicity is a law inherent in the "substratum of events," a law which breaks "through into phenomena otherwise regulated by the despotism of the mind." (pp. 198–199)

Had Eddington been fully determined to develop a concept of reality consistent with the operational attitude, he would have had to construe reality solely as the synthesis of all possible points of view, making of the common ground shared by all perspectives no more than a perspective shared by various observers; but his frequent characterization of this common factor as independent of any point of view carries him over into the realist concept. In his early work, the distinction between the two concepts is not so pronounced, although it underlies many of the statements I have quoted from *Space, Time and Gravitation*. In his Gifford lectures, the distinction is more clearly stated, although he makes no more attempt to unite the two views than he had done in the earlier work. But in these lectures he draws the distinction between the relative and absolute motion of objects, a distinction which leaves no possibility of merging the realist into the constructivist concept of reality. Speaking of the necessity of specifying the frames of reference before we make assertions about events, he remarks:

Since the motion can equally well be described as a motion of ourselves relative to the object or of the object relative to ourselves, it cannot influence the absolute behavior of the object. The apparent changes in the length, mass, electric and magnetic fields, period of vibration, etc., are merely a change of reckoning introduced in passing from the frame in which the object is at rest to the frame in which the observer is at rest. (NPW, p. 62)

But the strongest evidence of the realist concept in Eddington's thought comes from his doctrine of time, where he insists that the internal feeling of becoming is more than symbolic in that it gives a "true mental insight into the physical condition which determines it." (NPW, p. 89) There is no image building or construction at work here: there is, it would seem, a direct, non-structural contact with reality. Eddington anticipated that objections to this doctrine would arise, since it "is tantamount to an admission that consciousness, looking out through a private door, can learn by direct insight an underlying character of the world which physical measurements do not betray." (NPW, p. 91)

He steps over his doctrine of structure, but the world he enters is the same realist world to which the perceptual doctrine of structure refers.

> That dynamic quality – that significance which makes a development from past to future reasonable and a development from future to past farcical – has to do much more than pull the trigger of a nerve. It is so welded into our consciousness that a moving on of time is a condition of consciousness. We have direct insight into 'becoming' which sweeps aside all symbolic knowledge as on an inferior plane. If I grasp the notion of existence because I myself exist, I grasp the notion of becoming because I myself become. (NPW, p. 97)

In making becoming the condition of consciousness, Eddington is close to Kant's *a priori* conditions, but he fails to make anything of this suggestion. He is concerned to drive home his conviction in the directness of our temporal knowledge. No inferential structural process is required to reveal the objective quality of reality in this case: intuition short-cuts mediate inference. He does not indicate what his reasons are for this belief other than the fact that he believes the real world must be dynamic. He admits that the purely subjective account of becoming would "be adequate to account for the observed phenomena. The objections to it hinge on the fact that it leaves the external world without any dynamic quality intrinsic to it." (NPW, p. 92)

He reacted against the implications of his own subjectivist doctrine as applied to the temporal series; for, while he was content to allow the spatial characteristics to be products of our tools of measurement and hence a quality only of the subject-object relation, he did not like to place the temporal characteristics on the same basis. Time as perceived is an intrinsic property of reality, regardless of the perceiver. Time as it functions in the experiments of science is, like space, a product of scientific concepts and measurements, but "the relativity theory is not concerned to deny the possibility of an absolute time." (STG, p. 163) A philosophical interpretation of science must, Eddington believed, emphasize the absolute nature of time, even though this time is not directly contained in the operations of science. A basic belief of common sense in the dynamic quality of reality usurps the discussion of the relativistic nature of time. Common sense is used as a guide for interpreting the scientific data of time in a way congenial to this basic belief. We are confronted here with another category not mentioned by Eddington, a category of a unique sort. The category of the dynamic character of reality, of the reality of becoming, does not force our perceptions into a definite mould, but forces instead our interpretation of scientific data into harmony with common sense. The mind receives temporal data in two ways: as the editor of the sensible cryptogram and directly as a participant in time.

> We picture the mind like an editor in his sanctum receiving through the nerves scrappy messages from all over the outside world, and making a story of them with, I fear, a good deal of editorial invention. Like other physical quantities time enters in that way as a particular measurable relation between events in the outside world; but it comes in without its arrow. In addition our editor himself experiences a time in his consciousness – the temporal relation along his own track through the world. This experience is immediate, not a message from outside, but the editor realizes that what he is experiencing is equivalent to the time described in the messages. Now consciousness declares that this private time possesses an arrow, and so gives a hint to search further for the missing arrow among the messages. (NPW, p. 100)

Under this stimulus the editor-mind discovers in the concept of entropy a close parallel to the direct awareness of becoming, experienced as a traveller through time. "Entropy-gradient is then the direct equivalent of the time of consciousness in both its aspects." (*Ibid.*, p. 101) The concept of reality is, on this point clearly controlled by the naive realist belief of common sense. Not only is the operational concept diverted into a realist position, but the dualism of the structural concept is also abandoned. It is difficult to determine just why Eddington felt that the reality of becoming could be preserved only on a direct realist position, especially after the careful development of his perceptual doctrine of structure, for a dynamic quality can be predicated of reality without subscribing to Eddington's intuitionist position. Dobbs has recently developed some of the details of such a theory within the context of a dualist ontology.

> A 'psycho-neural parallelism,' or one-one correspondence, is postulated between features of certain 'experiential events' (namely, those experiential events normally held to be happening to some person's *mind*, which are describable in the language of psychology); and features of certain 'physical events' (namely, those events described in the language of physics, chemistry and physiology, which are ordinarily conceived as happenings in that same person's *body*). These physical events are conceived of as being causally connected with events in the physical world ourside the experient's body, by means of the concepts of light waves,

sound waves, chemical stimuli, and consequential processes in the nervous system (central and peripheral) and sense-organs, in the usual way.[1]

The isomorphism of Köhler is united to a transcendental doctrine of structure which connects the neural events of the brain with the external world of the dualist theory of perception and ontology. Dobbs does not make the doctrine of structure as it applies to this level any more specific or clear than did Eddington or Russell; but, since Eddington did accept this doctrine and give it such a place of importance in his philosophy, it is strange that he felt it necessary to transgress his basic dualism by advocating a direct acquaintance with the temporal sequence of events in the external world. On either view, providing the doctrine of structure can be made intelligible, the dynamic character of reality would be preserved.

The constructivist concept is, of course, antithetical to the premilinary dichotomy he established in the Gifford lectures, the separation between the scientific and common sense worlds. The scientific table is an object of inference. It is never an actual or possible ingredient of any one's perspective, although Eddington would seem to hold that the temporal characteristics of the scientific table, just as much as of the common sense table, are objects of direct acquaintance. Inference, of course, infects the constructivist or phenomenalist concept, but only in one direction, within the area of phenomenal activities. The causal theory of perception with its transcendent inferences from sense-data to stimulus is incompatible with the constructivist concept of reality, but it is not necessarily incompatible with the subjectivism of Eddington's epistemological interpretation of physics. He may be correct in saying that "It is the inexorable law of our acquaintance with the external world that that which is presented for knowing becomes transformed in the process of knowing." (NPS, p. 7) Such a claim, however, has not been justified. But the data of sensation, however distorted they may be, are for Eddington still data of a world beyond the nerve endings of the observer. The solution of the cryptogram of sense is controlled to some extent by the natural categories of the mind, but also "by

[1] BJPS. *op. cit.*, p. 122. Dobbs does not develop the details of this temporal isomorphism, concentrating upon the problem of the specious present. But the view here outlined stands as the background against which his solution of this further problem is presented.

studying the *recurrency* of the signs and indications." (NPS, p. 8) The inference made to a common external world is not presented in *New Pathways in Science* as just the synthesis of various perspectives. The inferences, made necessary by the recognition that other minds possess similar data, lead us to "the conception of an external domain (physical space and time) to contain the inferential objects" of our perception. (p. 10) The physical objects in the world of perception are also in the world of each man's perception. "There *is* an external world not part of the mind of either of us, but neutral ground wherein is located the basis of that experience which we hold in common." (NPS, p. 323) In seeking to uncover the subjective element in the laws of nature, Eddington insisted at the same time that there are objective laws. "If we are to discern controlling laws of Nature not dictated by the mind it would seem necessary to escape as far as possible from the cut-and-dried framework into which the mind is so ready to force everything that it experiences." (NPW, p. 210) He actually distinguished three different types of laws: identical, statistical, and transcendental. Identical laws are those "obeyed as mathematical identities in virtue of the way in which the quantities obeying them are built." (NPW, p. 244) Statistical laws relate to the activity of groups or to the expectancy with regard to some characteristic of those groups.

Neither of these can be said to be genuine laws of control of the physical world, since both depend upon the observer and his tools of measurement. If there are any genuine laws of nature (and Eddington seems to feel there are), they must belong to the third class of transcendental laws, those which transcend the operational, phenomenalist world.

> It is a natural suggestion that the greater difficulty in elucidating the transcendental laws is due to the fact that we are no longer engaged in recovering from Nature what we have ourselves put into Nature, but are at last confronted with its own intrinsic system of government. (NPW, p. 245)

He was rather hesitant about accepting these implications of his realism, but he concluded that "it is perhaps as likely that after we have cleared away all the superadded laws which arise solely in our mode of apprehension of the world about us, there will be left an external world developing under genuine laws of control." (*Ibid.*) Eddington did not claim to have discovered any of the transcendental laws: he even disclaimed that the interval, the four-fold dimensionality, belonged to this transcendent realm, attempting, as we have seen, to assimilate them to the constructivist view. In *The Philosophy of Physical Science* he drew a distinction between laws of nature and special facts, arguing that the latter were objective features of the physical world. The special facts differ from the subjective laws of nature in that the former can and the latter cannot be predicted. Special facts are just patterns "gratuitously incorporated in the design of the universe." (PPS, p. 63) He argues that although it is difficult to isolate "an objective law as completely as a subjective law, since it would have to be presented to us *via* our subjective forms of thought," nevertheless we could at least "detect a regularity and recognize that its origin was objective, even if we could only describe it in subjective terms." (*Ibid.*, p. 69) However, he does not tell us how this distinction could be recognized, just as he nowhere meets the problem of justifying his claim of the existence of an external world. He merely offers the rather cryptic remark that life, consciousness and spirit are instances of special facts, having in mind doubtless his doctrine concerning the mental character of reality.

Whether reality be mental or physical, whether there is any difference in kind between the existence of object and subject, Eddington's realist concept of reality compelled him to distinguish the knowing subject from the 'ultimate operand' upon which the selective activities of the subject are directed. An objective peg is required for knowledge even though this has "disguised itself to resemble the cloaks" of actual knowledge. (PPS, p. 59) He admitted that "so far as physical theory is concerned, it makes no difference whether we *create* or whether we *select* the conditions which we study." (PPS, p. 110) Science can restrict itself to the operational area without considering these larger questions. But a philosophy of science would seem committed to dealing with these questions of knowledge and reality. Eddington cannot escape responsibility for discussing them, since he has explicitly accepted a realist concept of reality. The physical world is carefully defined as the structure of the external world (*Ibid.*, p. 150); but an essential weakness adheres to the realist concept in his system because he failed to provide adequate arguments supporting his claim. The distinction between laws and special facts, between objective and selective subjective features, presupposes the ability to make the distinction between subject and object. He admits that the line cannot be clearly drawn and falsely concludes from this fact that "we have the alarming thought that the physical analyst is an artist in disguise, weaving his imagination into everything." (*Ibid.*, p. 111) He failed to see that in the absence of a definite line between objective and subjective, both the phenomenalist and realist conclusions stand on equally plausible ground. The tools may exactly fit the universe. In *New Pathways in Science* Eddington places great value upon recurrent experiences as an indication of an external and independent world, arguing even that we would never "have made progress with the problem of inference from our sensory experience, and theoretical physics would never have originated, if it were not that certain regularities and recurrencies are noticeable in sensory experience." (p. 8) But the mere fact of recurrence does not make legitimate the conclusion that recurrence is evidence of a sender of messages or that the data that recur should be taken as signs or symbols. The whole theory of cryptograms of sense stands on the same tenuous footing. Recurrence could of course be applied to the phenomenalist world of perspectives, denoting the common, shared material of different observers. Such application may be all that Eddington wished to make of the recurrent data of experience; but whatever the application, the realist concept remains unproved. The real force behind Eddington's conviction in an external, independent world is to be found in his *a priori* doctrine. His procedure is similar to Kant's. He apparently did not feel constrained to produce a definite criterion for distinguishing between objective and subjective, since he had shown (so he thought) that certain matters of fact were deducible from his epistemological premises. Such deductions could only be made if our knowledge was, to that extent, subjective. Thus, he made the distinction between objective and subjective depend upon the hypothetical method of his *a priori* doctrine. If X, Y, and Z can be deduced from the methods and procedures of science they cannot be objective features. Similarly, if A, B, and C cannot be deduced (that is, if they are special facts), they must

be objective features of the external world. The doctrine of
structure, if vindicated, would provide sufficient material for this
crucial distinction, but that doctrine itself presupposes the very
distinction it is designed to vindicate. Even if deductions could be
made of the sort Eddington suggests, it could be argued that the
proper conclusion was not that these factors were subjective, but
rather that the isomorphism between thought and reality had
been demonstrated, the rationality of the real shown. But, had
Eddington succeeded in establishing his doctrine of *a priori*
deductions, he would have provided a good basis for his doctrine
of structure and a point from which he could have argued to the
existence of an external world, provided he called attention to
the fact that some features of human knowledge are not *a priori*
in his sense of that term. However, since we have seen that his *a
priori* doctrine reduces to the common claim of a deductive
formulation for the data of science, his realist concept of reality
is denied this Kantian foothold. He has not succeeded in supplying what a host of other realist philosophers have failed to
supply, a proof for their beliefs in an external world independent
of human knowledge. Good evidence can be cited in favor of
taking this belief as a fundamental category or standard of an
axiomatic character in the systems of Russell and Eddington.
Historical analysis of the dualist tradition in British philosophy
would reveal a similar function played by the realist concept at
work in the systems of the other men of this tradition.

Interpreted in terms of the distinction between special facts
and laws of nature, Eddington's system might be assimilated to
the constructivist concept of reality. Interval, order in four
dimensions, and even the geodesic curve of freely falling bodies
would then become factors of the world of appearance, whereas
the specific amount of length, duration, and force would vary with
reference frame. But Eddington's realist claim for time makes it
difficult to carry out the assimilation; for, if we were to merge
the felt process of becoming with the constructivist concept, we
would mean only that the feeling of becoming, experienced by
one individual, was similar to that experienced by another. The
realist claim for time implies more than this. The unfolding of
nature is not offered as a process of appearances, as a to-and-fro
of ingredients in human perspectives. The claim for a distortion

of reality is meaningless unless it is meant to refer to a world independent of perception. Most important of all, the doctrine of structure, especially in its non-mathematical form, would also be without significance if meant to apply only to the phenomenal world shared by all observers. Like Russell, Eddington invoked the doctrine of structure to account for the difficulties of making the causal theory fit into the framework of science. The symbolic character of scientific knowledge is required to be representative of the world of common sense. Strict naive realism being impossible for all ontological properties except time, both Russell and Eddington have sought to harmonize common sense with their doctrine of structure. The order which is so crucial for structural knowledge is not meant to be predicated of a common phenomenal world, but of the non-operationalist realm left after naive realism is assimilated to the dualism required by the separation between scientific and common sense worlds. It would almost be possible to interpret Eddington in two different ways, providing that we suppress certain assertions concerning time and take the doctrine of structure in its mathematical sense only. For then the structural knowledge could be assimilated to his constructivist concept of reality, since the structural knowledge would be the mathematical summary or scansion of groups in the shared perspectival worlds. But the analysis of the doctrine of structure has disclosed that even in *The Philosophy of Physical Science*, where the mathematical side of the theory is given its greatest emphasis, the theory is not divorced from the perceptual problem associated with the causal theory. The structural knowledge in that book is still knowledge "of that which is outside everyone's mind." (p. 142) The disturbance of nerve terminals is taken as the end product of a long causal chain involving the unsensed physiological factors of electrical and chemical changes as well as the more distant causal properties of the object. The cryptogram presented to the mind by sensation is cryptic, not in the sense of carrying coded messages pertaining to the common perspectives of other observers, but in the sense of containing meanings of non-sensible entities. The insights, furnished by our sense of time and by our structural knowledge, into the external world posited by common sense, are meant to be realistic insights. In his later periods, Eddington was concerned with elucidating the subjective character of scientific knowledge, with those generalizations of objects imposed by our categories and procedures; but even there he recognized the necessity of placing these subjective generalizations in their objective context. He distinguished between the operator and the operand and spoke of the "objective particles" as opposed to our knowledge of those particles. (PPS, p. 37) The world of common sense, turned from a naive realistic to a dualistic concept, is symbolized in the complex structures of scientific knowledge. The perceptual aspect of the doctrine of structure cannot be ignored. The realist concept of reality so prevalent in Eddington's thought cannot be assimilated to the phenomenalist or constructivist concept. For the realist tradition inherited by Eddington, the operational language formulates only one aspect of scientific knowledge. The dualism of his position demands a linguistic formulation appropriate to the dualistic side of this knowledge.

Operationalism, with its linguistic convenience for expressing the processes of science, was accepted by Eddington; but he saw the inadequacies of such a linguistic formulation for the whole of science. A clear statement of the precise relations between the operational-phenomenalist and the realist concepts of reality is not found in Eddington's writings. But we can see how the two sides of his thought could have been united. If we read Eddington as not having two theories of structure but only one, the perceptual one which he held in common with Russell, the mathematical doctrine fades into the background and serves the purpose of aiding in the exact formulation of certain aspects of science. We can then place the operationalist and constructivist strands of his thought along with the mathematical deductive side as constituting various means of formulating the results of science. The causal theory, the perceptual doctrine of structure, and the realist concept of reality then emerge as concepts pertaining not to the linguistic formulation but to the ontological and epistemological presuppositions of scientific knowledge. Like many of his philosophical contemporaries, Eddington was interested in linguistic problems, although he never labeled them as such; but unlike the current stress, he was also concerned with problems of reality. He was the inheritor of two streams of thought in British philosophy: the older, classical forms of empiricism which never allowed the interest in linguistic analysis to dominate the concern with ontological problems, and the newer forms of thought which stem from the former but which pretend to see in the analysis of language the way towards the solution of the onto-logical questions of the older philosophy. Eddington was not aware of the duality of his inheritance, but its traces are clearly evident in his philosophical interpretation of the physical sciences.

# CHAPTER VII

# LINGUISTIC AND EPISTEMOLOGICAL DUALISM

In the present century, there are two main approaches to problems of knowledge within philosophy. The first of these is non-psychological and concerns itself with logical justification of the knowing process. The contention is that a philosophical investigation into human knowledge must not invoke data from the empirical sciences, since these in turn presuppose perceptual knowledge as a means of acquiring their data. The concern is to arrange the parts of the process of knowledge into some sort of logical order such that from basic simples we can derive the complex levels of knowledge. Russell, Price, Broad and Eddington are representative of this approach although none of them, least of all Eddington, has concentrated solely upon the formalization of perceptual and ontological beliefs. Eddington's concern with formulating the conclusions of physics in a deductive system reflects the same interest in formalization as do the sense-datum analyses of perception offered by the other three. The second main approach, which I have termed organic phenomenalism in chapter II, is more outspoken in its reference to scientific data, insisting that its objective is not to present a logical ordering of knowledge at the risk of ignoring empirical data, but rather to incorporate the empirical psychological data into its analysis of knowledge. There are several components in this second approach to epistemology. All cognition is said to rest upon a pre-cognitive level of experience. The organism and its environment are interacting on this level of experience in such a way that the later cognitive discriminations are grounded in the neuro-physiological structure of the organism. Knowing, like learning, consists in making discriminations in the originally undifferentiated field of experience, a discrimination which is guided by the needs and desires of the organism. The process of objectification, of becoming aware of a world distinct from ourselves, arises out of the sensory-motor activity of the organism which classifies in a neuro-physiological manner the qualities of experience, forms them into groups controlled by relations like similarity and contiguity, and then, when cognitive awareness emerges, dissociates these grouped qualities from the organism and gives them an independent status. This act of grouping becomes one of the essential defining properties of intelligence.[1]

The detailed elaboration of this organic phenomenalist position yields a philosophical theory culled largely from psychology. It is best described as epistemologically phenomenalist but ontologically dualistic. Man's knowledge of his environment cannot be separated from the means at his disposal for apprehending and manipulating that world.

> The position here taken holds that, since every special case of knowledge is constituted as the outcome of some special inquiry, the conception of knowledge as such can only be a generalization of the properties discovered to belong to conclusions which are outcomes of inquiry.[2]

At the same time, the organic phenomenalist recognizes the reality of the environment as an independent ontological factor. Dualism and phenomenalism find a more easy union in this approach to problems of knowledge, primarily because the point of departure is taken as the intereaction of organism with environment. But it is not the dualism of this position which provides significant points of comparison with the position of Eddington; the role of phenomenalism in this approach does offer such comparison. When the concept of reality is found to be evolved from the bodily activities of movement and countermovement, the phenomenalist and operationalist attitudes become at once theories of reality and linguistic formulations of the concept of the real; for the real is intimately bound up with our movements and our operations. Sensory-motor activity becomes a practical expression or illustration of the operational concept of reality. The concept of the real for the organic phenomenalist

---

[1] Cf. Piaget's *La Construction du Réel chez l'Enfant*. For a further elaboration of the organic phenomenalist position, see my discussion, "Philosophical Realism and Psychological Data", *Philosophy and Phenomenological Research*, June 1959, and my review of Maurice Natanson's *The Social Dynamics of George H. Mead*, in *The Journal of Philosophy*, January 29, 1959, pp. 140–145.

[2] Dewey, John, *Logic, The Theory of Inquiry*, p. 8. It should be noted that what I have called organic phenomenalism is by no means limited to pragmatic theories of knowing. It finds its more interesting exemplifications in such neo-Kantian writers as Husserl and Cassirer.

is a concept defined in terms of the operations of the organism upon its environment. When phenomenalism and operationalism appear within the context of Eddington's dualistic acceptance of the causal theory of perception and the perceptual doctrine of structure, it ceases to be a theory of the real and becomes primarily a linguistic formulation for the data of science.

Eddington's operationalism combines with his mathematical theory of structure and the constructivist concept of reality to constitute the basic language in terms of which the data of science are formulated; it serves the same function as the basic proposition in Russell's linguistic analysis. Had Eddington concentrated upon the linguistic aspect of his philosophy of science, he might have seen the parallel between operationalism as it functions in his system and the basic language of Russell's *Inquiry into Meaning and Truth*. Operationalism is the basic language for the formulation of the data of the scientist but this language is inadequate for formulating the whole of Eddington's philosophical analysis of science. The referents of the sensory and intellectual cryptograms in the perceptual doctrine of structure cannot be formulated in this basic language. Phenomenal and trans-phenomenal emerge from his philosophy of science; but, though he recognized the importance of precise formulations within science, he was not aware of the need for a theory of meaning to explain the dualistic aspects of his general ontology, nor was he conscious of the relation between linguistic formulation and ontology. When we view the place and function of his operationalism in the total compass of his philosophy, we find that it functions as a linguistic formulation of only one side of his system. The inheritor of the two strands of British philosophy, linguistic and ontological analysis, Eddington quite unwittingly moulded his philosophy to fit this dual heritage, but the mould did not fit well, the two parts did not join smoothly. Even with Russell, the difficulties of uniting such disparate concepts as that of a hard core of indubitable data and the trans-phenomenal world have not been wholly overcome either epistemologically or linguistically. There is an important problem of formulating these two parts of the dualistic theories of knowledge and reality into an adequate language, a problem which has its ontological counterparts in the difficulty of making inferences from sense-data to physical objects. It is this problem which offers us a way of uniting the two sides of Eddington's system, although the union will not be an entirely happy one. We can learn from a close study of this problem just what the difficulties left unsolved by Eddington are and to what segment of his system they refer.[1] We can discover as well that the current craze in many philosophical quarters concerning the omniscience of language analysis is mistaken in at least one important area: analysis of language does not yield a solution to the central problem of dualistic theories of knowledge.

In brief, the linguistic problem which arises on a dualistic analysis of our knowledge of the external world, is as follows. It is recognized that many of our beliefs about the external world contain assumptions and convictions which are not revealed in any one experience, or even in any totality of sense experiences. Thus, a distinction is thought to be necessary between the given and the inferred elements of such beliefs on the assumption that whatever is found out is found out *either* by inference *or* by being something given. The basic epistemological problem issuing from this distinction is: what will justify our passage from the given data of our sense experiences to the inferred and accepted objects of our perceptual experiences? Whatever the answer be, we must be able to state it in language which will not itself cloud the nature of the problem by failing to distinguish between the given and the inferred. We require one formulation for sense-datum experiences and one for perceptual experiences. We need as well statements showing or justifying the validity of making the transition from the first set of statements to the second. In other words, when we seek to verbalize the answer to the dualistic problem, we meet the same kind of issue on the linguistic as on the epistemological level. Can we construct statements of the two required kinds, and so construct them that we will be able to provide for a verbalization of the epistemological transition from sense-data to physical objects? A necessary preliminary to answering these questions and to suggesting a way of uniting the two sides of Eddington's philosophy of science is a clear understanding of the epistemological role of basic propositions in a dualistic theory of knowledge.

---

[1] Pp. 130–141 contain a revised form of an article appearing in *Mind*, January 1953 under the title of this chapter.

Russell provides, in his *Inquiry into Meaning and Truth*, a convenient analysis of this problem, his task being to construct a language which will fit the deductive scheme which he feels is requisite for precise knowledge. Just as Eddington has sought to formulate the data of the physical sciences in a deductive system, so Russell has tried to find a deductive pattern of premises and conclusions which will be adequate to state the general epistemological processes of knowledge. Instead of constructing a specific body of exact data into a deductive inference system and its corresponding linguistic correlate, Russell's task is to work from the data of perception and to construct a system of linguistics and of inferences which will satisfy the dualistic analysis to which he adheres. The pattern of deduction is the same but the contents of the inferences and of the premises have been altered.

In general, Russell considers all epistemological premises as having three main characteristics. They must be logical, psychological, and true, so far as we can ascertain. These propositions permit of being ordered in such a manner that a deductive system can be based upon them. It is the inferential order provided by the deductive arrangement of such propositions or premises that Russell calls their 'logical' character. The logician, in other words, seeks for that minimum set of premises from which he can deduce the rest of the propositions in a given system. Eddington and Milne sought to apply this scheme to modern physics. Russell, as an epistemic logician, attempts to apply this inferential order to epistemology; but he does not intend the deduction of epistemological propositions to be made solely from basic propositions, those propositions constituting the axiomatic material of epistemological premises. The deduction will be made, if at all, from the total set of epistemological premises, which is much broader than basic propositions. Within this range there must be certain primary propositions which form the epistemological, not the logical, basis for the other propositions of the class. Russell does not say or mean to imply that deductive inferences can be made from one epistemological premise to another: the deduction is from epistemological *premise* to *conclusion*. It is Russell's conviction that certain of these premises must be basic in order to make valid inferences from the class of epistemological premises.

Eddington's attempts to uncover categories of thought and to establish an epistemological *a priori*, are instances of this search for basic axioms from which to establish valid deductions. But whereas Eddington sought to use these epistemological premises for concrete inferences to the laws of nature, Russell is concerned only to use them as the grounds for formulating human knowledge in general into a deductive pattern. From the psychological side, such basic epistemological premises can be defined as those beliefs which are not caused by any other beliefs.

> Psychologically, any belief may be considered to be inferred when it is caused by other beliefs, however invalid the inference may be for logic. The most obvious class of beliefs not caused by other beliefs are those that result directly from perception.[1]

But it is not the beliefs which *actually* do result from perception that Russell is concerned with, it is those which *should* result, epistemologically. When it functions as a premise for epistemology, such a fundamental belief is called a 'perceptive premise.' It is necessary to define rigidly the meaning of 'perceptive premise,' since a belief, such as that expressed by the statement 'there is an eclipse,' might be taken for a perceptive premise; but such a belief actually "goes beyond the mere expression of what I see" and hence cannot be valid for the sense-datum theory which Russell advocates. Epistemology, in other words, demands a 'perceptive premise' "which there is never good reason to think false, or, what comes to the same thing, something so defined that two perceptive premises cannot contradict each other." (p. 135) Russell insists that what he calls 'momentary empiricism' has to be the starting point for all empirical theories of knowledge, since it alone starts from that which I can know without dependence upon anything other than my perceptive experiences. Eddington, as we have seen, followed Russell in this conviction of starting with momentary empiricism. Since such premises are going to form the basis for all other propositions in the dualistic theory of knowledge, it is essential that they be independent of other beliefs and other propositions. They constitute the ground of later inferences and hence cannot themselves contain inferences. Basic pro-

---

[1] *Inquiry into Meaning and Truth*, p. 132. All further references in this chapter to Russell are to this work, unless otherwise specified.

positions are thus perceptive premises which are a "sub-class of epistemological premises." They do not constitute the sum of necessary premises for an empirical epistemology, but they form the basis of the system. Such propositions must be known independently of any evidence, "since there must be a perceptive occurrence which gives the cause and is considered to give the reason for believing the basic proposition." (p. 138) One of the questions which emerges from the relation of basic propositions to perceptive experiences is "what do we know when we know that our words 'express' something we see?" That is, "when you see a black object and say 'this is black,' you are not, as a rule, noticing that you say these words; you know the thing is black, but you do not know that you say it is." (p. 60) However, when we are, as Russell is in the *Inquiry*, "studying the relation of language to other facts," we do take notice of a connection between the words and the non-verbal facts we are seeking to express. The nature of this connection is an important epistemological question. On the level of basic propositions, the relation must be such that no other basic proposition can contradict them. These basic beliefs and their linguistic translations must be restricted to that which can be known from sensation alone. We must avoid what Russell calls 'condensed inductions,' which go beyond the given or the sensational core, since these are not basic in the sense of being free from inferences.

The basic propositions which Russell is seeking are propositions which assert what is given in experience as distinguished from that which is interpreted or that which is inferred. But there has been a confusion over what is meant by 'the given.' Eddington suggested that these primitive, given data, though spoiled by intellectual activities, consisted of sensations, emotions, conceptions, and memories. In the theories of Russell and Price, the given has been more concisely restricted to sensings of the moment, but all those advocating the doctrine have admitted that the given was probably an intellectualistic abstraction from the actual processes of perception. Modern psychology has pointed out that we do not perceive discrete sense-data, but complete forms, already integrated into definite patterns. Qualities never come to us separately and alone but always as the qualities *of* some object.

Whatever the manner of expression, the phenomenological fact is simply that in perception we are *conscious*, in one sense of the word, of physical objects, without at the same time being conscious, in another sense of the word, of the entities which have traditionally been called 'sense-data.' Perception, in short, is not a twofold state; and since we *are* conscious of physical objects we cannot possibly be conscious of sense-data in the distinctive manner required by the Sense-Datum Theory.[1]

Eddington was not concerned to enter into the various philosophical disputes about the nature of the given; but his position commits him, as we saw in chapter II, to many of the same problems characteristic of these debates. In his case, as in that of Russell, the insistence that perception starts from sense-data is not meant to negate the findings of gestalt psychology. Even if Eddington was unaware of the claims of these psychologists about the nature of the perceptual act, his agreement with Russell concerning the genesis of perception need not be taken as a violation of these psychological findings; for the contemporary British epistemologists who advocate the sense-datum analysis do not offer it as an accurate description of the actual process of perception. It is true to say that none of the sense-datum philosophers have related their claims to those of the psychologists (and philosophers) who offer empirical evidence showing that perception takes place by awareness of wholes rather than discrete sense-data. As a result, there is no clear statement in the literature of what I should take to be the basic ground of the sense-datum dualistic theory. The given which is so much appealed to in these theories is what I should call the *epistemic given*. Sense-data are not the beginnings of perception psychologically and very probably never function in our actual awareness of the world, but they are the analytic beginnings, in the sense of being products of logical analysis, which are demanded for epistemology by two non-epistemological factors: a specific methodology and a particular ontology.

The methodology which operates in most contemporary forms of epistemological dualism is well known, not being limited to procedures in epistemology. It consists of passing from basic axioms which do not require proof to propositions, assertions, or beliefs which are the products of inference. The rationalists have sought to use this method in all spheres of thought and it is the

[1] Firth, R. "Sense-Data and the Percept Theory," *Mind*, vol. 58, 1949, p. 449.

nerve of the geometric method so much practiced by Eddington and Milne. The sense-datum philosophers have adopted this methodology for the purposes of epistemology, some less critical than others of its nature. Preference for this specific methodology led these philosophers to seek for indubitable factors in perception. That preference seems to be one of the dominant causes for the tenacity with which adherents of the sense-datum theory cling to their position. However, there is another more important presupposition to the theory, a factor which is less explicitly present in the writings of sense-datum philosophers, but one which has made the theory appealing even in the face of the factual refutations cited by many writers. This factor is the ontological position referred to in chapter III: the assumptions concerning the nature of physical objects. All the sense-datum philosophers begin their analyses explicitly or implicitly with a definition or description of the nature of the ontological objects which are the referents of perception, a description which itself assumes ontological dualism. Broad lists five characteristics of such ontological objects which he believes are pervasively accepted, even by those outside the sense-datum theory.[1] By dividing the object into those aspects common to the various phenomenal fields and those which belong to a trans-phenomenal field, the dualism implicit in Eddington's system mirrors that of his philosophical contemporaries. When the ontological assumptions concerning the nature of reality are related to the methodology in common use among the sense-datum philosophers, we find the commitments to dualism are made even stronger. Methodology and ontology demand a particular epistemology. Thus, the given is epistemic and not psychological. Recognition of this distinction makes clear certain of the fundamental problems facing the linguistic formulation of a system such as Eddington has adopted. For now the question raised by Russell, 'what linguistic form will adequately describe perceptive experiences?,' can be seen to be meant in the epistemic sense: the formulation must be adequate to the epistemic given. Perceptive experiences are the experiences which basic propositions are said by Russell to describe. Consequently any description or translation of them must not go beyond the epistemic given in its linguistic expression. The total set of verbal proposi-

[1] See above, pp. 27–28.

tions which describe perceptive experiences in this way constitutes a large and important part of that class of statements which Russell calls the 'object language,' or alternatively, the 'primary language.' He defines this object language as that language which consists wholly "of object words, where 'object words' are defined, logically, as words having meaning in isolation, and psychologically, as words which have been learnt without its being necessary to have previously learnt any other words." (p. 65) These are words which, unlike logical ones, do not depend upon a context for their meaning. This meaning is learned by confrontation with objects which are what they mean or instances of what they mean. (p. 26)

The object which is confronted is not always of the same type, for Russell explains that we can learn by confrontation, the names of people; class names, such as 'man' and 'god'; names of sensible qualities; names of actions; and relation words. Later in the discussion he makes it clear that words like 'dog' and 'cat' are also object words. The conjunction of physical object and sense-datum words in the same list of words learned by confrontation is significant in that it shows that in his discussion of the object language Russell does not always keep before his attention the epistemological considerations which in other places seem to be directing his argument. In the presentation of what he means by the object language, he has not been concerned to introduce his causal theory of perception; but in the above list of objects which are said to give rise to object words, Russell could not be asserting that this list is consistent with the causal theory without some important qualifications. Basic propositions are a very important aspect of the object language, and the considerations of the causal nature of perception are quite relevant to a discussion of basic propositions, since it is their linguistic function to verbalize certain of our perceptive experiences. Thus, it is necessary to know just what we can be said to know in such perceptive experiences. Clearly, we can be said to know two radically distinct kinds of objects. We know both operational-phenomenalist objects and realist objects. We know both of Eddington's two tables. We have already seen that that aspect of the object language which contains basic propositions must be epistemologically and not psychologically primary; if so, many of the object words which arise from the confrontation of such objects as Russell lists above will have to be dismissed on the grounds that they contain condensed inductions, and hence go beyond the immediate epistemic given. The knowledge which we are said to have of the scientific table is inferential and filled with condensed inductions. It cannot satisfy the requirements of the epistemic given and cannot function as the referent of basic propositions. On the purely descriptive level, there are many situations in our experiences which acquire names through the association of object with verbal utterance. On this level,

All that is essential to an object-word is some similarity among a set of phenomena, which is sufficiently striking for an association to be established between instances of the set and instances of the word for the set, the method of establishing the association being that, for some time, the word is frequently heard when a member of the set is seen. (p. 72)

But we acquire many object words in this way which do not stand the epistemological test: they are not restricted to what is epistemologically given in perceptive experiences. Russell is not unmindful of this fact. He does not mean to construct an object language which consists only of those words which are free from condensed inductions. The object language is much more comprehensive; but, what is most important, its foundations must consist of object words which are free from all inferred factors in order for those sentences using induced words to be epistemologically acceptable. Thus, it is the concept of the basic linguistic proposition to which Russell appeals, in order to escape such inconsistencies in his object language. There is a hierarchy of sentences within the object language, some being more basic than others in virtue of their fulfilling the requirements of the epistemic given. But how do words take on their trans-phenomenal or inductive character? According to Russell's analysis of the meaning of object words, it is established through the conjunction of object with words, heard, spoken, or read. In certain situations we have learned to associate the word 'dog' with a particular object in our experiences; in other situations, we come to attach words like 'red' to objects in our experiences. What is it in each of these situations which is named? Unless it can be shown how the name comes to mean more than is immediately presented in the epistemological sense, it would seem that all that is named in each case is a certain configuration of our perceptual field. Similar configurations come to have associated with them the same sounds or verbal tags. But if perceptual configurations exhaust the meaning of object words, the dualistic distinction between epistemological and ontological objects would never arise and we would thus be constricted within the operational-phenomenalist area. We can, of course, make a distinction between given and inferred factors on a purely phenomenalistic interpretation, since the whole of any phenomenalistic object is never given in any one experience. Inference must be made through the use of memory and past experiences to further obtainable sense-data. But it is not this kind of inference which is characteristic of dualistic epistemologies: vertical, transcendent inferences from sense-data to physical objects are required as well. The word 'dog' contains a condensed induction for Russell in the memory sense, but it also goes beyond experience in asserting the existence of an ontological object containing factors which are never directly sensed. These inferences from sense-data to the non-sensible factors raise a fundamental epistemological and linguistic difficulty. On his theory of the meaning of object words, Russell is unable to explain how this distinction enters our thought. Without it there is no difference between the psychological and the epistemic given; but the whole purport of Russell's discussion of epistemological premises and their linguistic form is that in translating from internal beliefs to linguistic statements, we must be constantly on our guard not to include references to that which is not epistemologically given. In other words, he assumes from the very beginning the above distinction which his theory of meaning seems unable to explain.

The same difficulty is present in a slightly different form in Russell's elaboration of the nature of the object language. He insists that the linguistic form of object sentences must be what he calls 'atomic.' He defines 'atomic' in several different ways. In the chapter where he first introduces this concept, he says "that a *form* of proposition is atomic if the fact that a proposition is of this form does not logically imply that it is a structure composed of subordinate propositions." (p. 34) What is single grammatically is not necessarily single or atomic in this sense. In another place, he defines atomic sentences more precisely as follows:

A sentence is of atomic form when it contains no logical words and no subordinate sentences. ... Positively, a sentence is of atomic form if it contains one relation-word (which may be a predicate) and the smallest number of other words required to form a sentence. (p. 95)

Neither of these definitions tells us all we need to know about the words which fill out the basic atomic propositions. Russell cannot mean to say, in the case of basic atomic sentences, that we fill in these sentences with any object word, since not all object words qualify for filling the requirements of epistemology for basic object words. In the chapter on proper names, Russell constructs a language which consists of relation words and names. He purports to give a syntactical definition of 'name' as any word which can occur in an atomic sentence. In the case of basic atomic sentences, however, the names must not extend, in their meaning, beyond the epistemic given. But the syntactical definition does not in itself tell us what sort of words can occur in the basic atomic sentences. What are the criteria for determining the proper sentential constituents of basic atomic sentences? The answer is that the criteria lie in certain epistemological considerations which are paramount throughout the *Inquiry*. The words which can occur in basic atomic sentences are only those which are restricted in their meaning and import to what is epistemically given in momentary experiences, an epistemic given controlled by the two non-epistemological premises of methodology and ontology. Thus, I would suggest that Russell has given an epistemological rather than a syntactical definition of names, as applied to basic atomic sentences. Syntactical considerations are subservient to epistemological requirements. How the epistemological requirements arise Russell leaves unexplained; but, if we are permitted to draw implications from his other writings, I think we can say that these epistemological requirements, in the form of the criteria of the meaning of physical object words, are presuppositions throughout Russell's epistemological discussions. Like Price and Broad, Russell assumes that physical object words contain a meaning-content which is never completely exemplified in experience. The implicit meanings for such words in Russell's *Problems of Philosophy* were that physical objects cannot change as rapidly or as frequently as our sense-data and that physical objects are multiply accessible to different people and by different sense-modalities. Moreover, like Price's physical occupant and Broad's scientific object, Russell posits the atomic structure of physics as an important non-sensible ingredient in the physical world. These meanings are inconsistent with his theory of meaning set forth in the *Inquiry*, since there is no way in which they can be accounted for on the basis of confrontation. However, they are necessary to lend significance to the central problem which his and Eddington's systems pose: the construction of a language which will express our basic, perceptive beliefs without transcending the epistemic given. The epistemological meanings which Russell implicitly assumes throughout with reference to such words as 'dog', 'apple', etc., contain condensed inductions: they transcend the sensible qualities within experience both in time and in space, referring to the physical world distinct from sensible qualities.

The need for this restriction in basic object words is very obvious in the chapter in the *Inquiry* on proper names, where Russell makes it clear that the words which are to appear in atomic sentences can only be words which do not transcend the epistemic given. The only words which conform to this condition are the names of sensible qualities or sense-datum words. Russell is concerned in that chapter to work out a linguistic formulation which will express all that we can safely express concerning our experiences, without going into the inferred world of mediated experience. Such a restriction indicates that he has unconsciously limited himself to the basic aspect of the object language. But his attempts to find a language which will be primary in the epistemological sense leads him into a phenomenalism which is strangely opposed to his causal theory of perception, a phenomenalism which comes close to splitting Russell's system into the same two sections as we have found in Eddington's sytem. The phenomenalism of his theory of meaning walks strangely in hand with his causal theory of perception. One of the merits of Russell's basic atomic sentences is that they enable us, he believes, to use a language which is free from all epistemic doubt. In the common sense language, statements of the form 'this is red' frequently occur, but Russell feels that the force of the word 'this' extends beyond the epistemic given and hence should be eliminated from our basic language. When 'this is red' is considered as a subject-predicate proposition, the 'this' "becomes a substance, an un-knowable something in which predicates inhere, but which, never-theless, is not identical with the sum of its predicates." (p. 97) Russell's concern to eliminate substance words like 'this' suggests that he was advocating a phenomenalism, but the denial of the reduction of the physical world to sensible qualities is entailed in his own causal theory. However, not being concerned with the origin of sensible qualities or with their status in the physical world but with the construction of a language which will serve as the basic, primary language for epistemology, Russell (in the *Inquiry*) is forced to remove the word 'this' from the basic object language for the same reasons that 'dog' is excluded. The state-ment 'this is red' must be changed into the form 'redness is here,' where 'red' is a name and not a predicate. When 'red' is considered as a predicate, the introduction into our language of references to a Lockean substance which is unexperienced seems inevitable. On the other hand, when we consider 'red' as a name designating some portion of continuous space-time in our immediate ex-perience, we need not introduce the concept of something over and above the sense quality. The dichotomy between subject and its qualities is thus avoided in our basic language. But unless Russell makes the further qualification that this is a language about epistemological objects only and not about the external world of ontological objects, he is involved in a position which contradicts his own causal theory. In the manner of Eddington, Russell was concerned with only one aspect of the problem of meaning within a dualistic epistemology. The basic object language, fulfilling the conditions of the epistemic given, is concerned with the same category of meaning as Eddington's operationalism. Russell was more aware of the need for a precise linguistic formulation of his epistemology than was Eddington, but both men overlooked the need for explaining and formulating the genesis of physical object words. They both adopted the commonly accepted meanings of these words without raising the question of their origins. It was these meanings which they were both trying to save in the face of all the refutations of naive realism. They were concerned to construct a theory of knowledge which would preserve as much of common sense as can possibly be made consistent with the facts, as well as with physics. They did not question the fundamental beliefs of common sense except as these conflicted with each other or with the findings of science. The dualistic epistemologies of Russell and Eddington were constructions within this framework. Failure to be critical of the accepted common sense definitions of reality led to the one-sided theory of meaning explicit in Russell's and underlying much of the analysis of Eddington's writings. The difficulty for Eddington lies in fitting operationalism into his objective realism. For Russell the problem takes the form of how to bring the physical object meanings presupposed in his causal theory into harmony with the phenomenalist criterion of meaning laid down in the *Inquiry*.

We have seen how they both sought for this needed harmony by means of the doctrine of structure, a doctrine which if true would enable us to infer from our sense-datum statements something of the nature of the physical object lying beyond the realm of direct perception. But the doctrine of structure is at best a speculative theory when applied to the trans-phenomenal realm. Perhaps because of his awareness of the tentative nature of this doctrine, Eddington sought further support for the unification of his epistemology in his doctrine of categories, ingrained concepts which mould the common sense as well as the scientific picture of the world. The way of return to common sense from the complex and abstract constructions of modern physics lay, for Eddington, along the route of these controlling categories. The nature of thought or of mind being as it is, it is impossible for the scientists to stray too far from the common sense picture. But the analysis of Russell's attempt to construct a satisfactory formulation of his dualistic system has disclosed that Eddington omitted from the realm of categories certain other fundamental concepts. The criteria of physicality implicit from the beginning in the theories of Russell and Eddington constitute a primitive orientation of thought. They form the ontological presuppositions in terms of which our very concept of reality is interpreted. With this criterion, the dualism of their epistemologies follows easily. In terms of linguistics, Russell needs to add as basic axioms of his system not only the doctrine of structure, the isomorphism of the method of correspondence, but like Eddington he must invoke his ontological concept of reality as being axiomatic as well. Without statements formulating these criteria in linguistic terms, the two sides of his system cannot be united. The phenomenalism of his *Inquiry* stands in sharp conflict with the dualism of his causal theory. The interpretation and reduction of physical object statements to sense-datum statements must be augmented by statements specifying that the transcendent meanings of certain object words are introduced as basic premises by way of saving as much of common sense as possible. Linguistic analysis is not sufficient by itself: ontological considerations must play their rôle. Concentration upon linguistic analysis leads to the pure operationalist position advocated by Dingle or to the phenomenalist perspective of Russell's *Inquiry*. The two concepts of reality which we have found at work in Eddington's philosophical interpretation of the physical sciences are in the last analysis related by way of genus and species. The realist concept is found to be the more pervasive and more active of the two, controlling all else. The operationalist concept turns out to be a linguistic formulation of only one side of the total picture in terms of which modern science needs to be interpreted. Milne may be correct in urging that an understanding of the world arises only when we can fit the empirical data to a rational scheme, but in the systems of Russell and Eddington, such understanding is dependent as well upon recognition of the basic concept of reality inherent in their approach. The constructivist concept of reality functions as a base for the linguistic expression of an important aspect of this concept of reality; but the implicit transcendent theory of meaning invoked by both men leads to a radical alteration of the immanent restrictions of their linguistic formulations. The separation between the operational-phenomenalist and the dualistic sides of their epistemology cannot be solved until full recognition has been given to the dualistic criteria of physical object words. By recognizing the dualist criteria of physicality as a basic axiom in their thought, we can add some sort of unity to the two sides of their systems. Without this recognition, both the linguistic expression and the epistemological doctrines of Russell and Eddington remain doomed to a vicious dichotomy.

Recent philosophy has shown a reluctance to deal with ontological questions, favoring a linguistic analysis of concepts. The prevailing form of philosophy of science is no exception to this general trend. But not only is it wrong to say Eddington was confused in failing to eliminate all traces of the dualist concept of reality; it is important to reinstate the philosophical legitimacy of non-phenomenalist ontologies. The failure of Russell and Eddington to solve the problems of dualism, in no way tells against it. Any analysis of Eddington's philosophy of science must recognize the twin strands of operationalism and dualism in his thought and must appreciate the possibility of uniting them along the lines I have suggested. What is required to make the union more smooth and acceptable is a careful analysis of dualist meaning. Linguistic analysis can perform a useful service, but it is time that philosophers recognize the priority of ontological requirements. If we wish to have from the philosophy of science more than a logical or semantical formalization of science, we must return to the ontological problems which used to excite us. A robust and mature philosophy of science can only emerge after we view science in the context of the traditional philosophical problems. A consideration of Eddington's writings can perform a useful function in pointing up again the possibility of such an approach to the philosophy of science.

# BIBLIOGRAPHY

Bergson, Henri, *Matière et Mémoire*. 46. éd. Paris, Presses Universitaires de France, 1946 c1939.

Born, Max, *Atomic Physics*. 4th ed. London, Blackie, 1946.

Boyle, Robert, *The Origin of Forms and Qualities*. London, 1666.

Bridgman, P. W., *The Nature of Physical Theory*. Princeton, University Press. 1936.

– *The Logic of Modern Physics*. London, Macmillan, 1937.

– "The Nature of Some of Our Physical Concepts," in *The British Journal for the Philosophy of Science*, 1951, I, pp. 257–273.

Broad, C. D., *The Mind and Its Place in Nature*. London, Routledge, Kegan Paul, 1925. (Tarner lectures at Cambridge for 1923).

– *Scientific Thought*. London, Routledge, Kegan Paul, 1923.

– Review of Eddington's *The Philosophy of Physical Science*, in *Philosophy*, vol. XV, 1940, p. 301–312.

Brown, G. B., "A New Treatment of the Theory of Dimensions," in *Proceedings of the Physical Society*, vol. 53, 1941, p. 418–431.

– "Why Do Archimedes and Eddington Both Get 1079 For the Total Number of Particles in the Universe?", in *Philosophy*, vol XV, 1940, p. 269–285.

– A Letter in *Nature*, vol. 148, 1941, p. 504.

Burt, C., *The Factors of the Mind*, London, Univ. of London Press. 1940.

Campbell, N., A Letter in *Nature*, vol. 139, 1937, p. 1005.

Cassirer, E., *Das Erkenntnisproblem in die Philosophie und Wissenschaft der Neueren Zeit*. Berlin, B. Cassirer, 1922–23. 3 vols.

– *The Problem of Knowledge, Philosophy, Science, and History Since Hegel*. New Haven, Yale University Press, 1950. (Especially Part I, "Exact Science," p. 21–118)

– *Substance and Function and Einstein's Theory of Relativity*. Authorized Translation by W. C. and M. C. Swabey. La Salle, Ill., Open Court pub. co., 1923.

Collingwood, R. G., *An Essay on Metaphysics*. Oxford, University Press, 1940. (Especially Part I and Part IIIb).

– *An Essay on Philosophical Method*. Oxford, University Press, 1933. (Especially Part VI,"Philosophy as Categoreal Thinking", p. 117–135)

Destouches, J. L., *Essai sur la Forme Générale des Théories Physiques*. Paris, 1938.

– *Principes Fondamentaux de Physique Théoriqeu*. Paris, Hermann, 1942.

Destouches-Février, Mme. P., "Monde Sensible et Monde Atomique," in *Theoria*, vol. XV, 1949, p. 78–90.

Dewey, John., *Logic, The Theory of Inquiry*. New York, H. Holt, 1938.

Dingle, H., *Through Science to Philosophy*. Oxford, University Press, 1937.

– "Modern Aristotelianism," in *Nature*, vol. 139, 1937, p. 734–786.

– "A Theory of Measurement," in *British Journal for the Philosophy of Science*, vol. I, 1950, p. 5–27.

– Letter in *Nature*, vol. 148, 1941, p. 341–342; 503–505.

– Obituary of Eddington, *Proceedings of the Physical Society*, vol. 47, 1947, p. 244–249.

– A Review of Eddington's *The Philosophy of Physical Science*, in *The Observatory*, vol. 63, 1940, p. 18–25.

- A Review of Arthur Pap's *The A Priori in Physical Theory*, in *Proceedings of the Physical Society*, vol. 60, 1948, p. 599–600.
- A Review of Russell's *Physics and Experiences*, in *Proceedings of the Physical Society*, vol. 59, 1947, p. 509–511.

Dobbs, H. A. G., "The Relation Between the Time of Psychology and the Time of Physics," in *British Journal for the Philosophy of Science*, vol. II, 1951, p. 122–142, 177–193.

Douglas, A. Vibert, *A. S. Eddington*. Edinburgh, F. Nelson and Sons, 1956.

Eddington, A. S., *Fundamental Theory*, ed. by E. T. Whittaker. Cambridge, University Press, 1946. (Posthumous).
- *The Mathematical Theory of Relativity*. 2d ed. Cambridge, University Press, 1924.
- *The Nature of the Physical World*. Cambridge, University Press. 1948 [i.e., 1928].
- *New Pathways in Science*. Cambridge, University Press, 1935. (The Messenger Lectures for 1934).
- *The Philosophy of Physical Science*. Cambridge, University Press, 1941. (Referred to in this study as his last work, meaning last published work in his life time).
- *Relativity Theory of Protons and Electrons*. Cambridge, University Press, 1936.
- *Science and the Unseen World*. Cambridge, University Press, 1929.
- *Space, Time, and Gravitation, An Outline of the General Relativity*. Cambridge, University Press, 1920.
- *The Theory of Relativity and Its Influence on Scientific Thought*. Oxford, University Press, 1922. (Romanes Lecture at Oxford)
- "The Meaning of Matter and the Laws of Nature According to the Theory of Relativity," in *Mind*, vol. 29, 1920, p. 145–155.
- "The Philosophical Aspect of the Theory of Relativity," in *Mind*, vol. 29, 1920, p. 415–421.
- A Letter in *Nature*, vol. 139, 1937, p. 1000–1010; A letter in *Nature*, vol. 148, 1941, p. 141, 236.

Einstein, Albert., "Die Grundlage der Allgemeinen Relativitätstheorie," in *Annalen der Physik*, vol. 49, 1916, p. 772–822.

Emmet, Dorothy M., *The Nature of Metaphysical Thinking*. London, Macmillan, 1945.

Firth, R., "Sense-Data and the Percept Theory," in *Mind* vol. 58, 1949, p. 435–466.

Gonseth, F., *Les Mathématiques et la Réalité, Essai sur la Méthode Axiomatique*. Paris, F. Alcan, 1936.

Gurwitsch, Aron., "Gelb-Goldstein's Concept of 'Concrete' and 'Categorial' Attitude and the Phenomenology of Ideation," in *Pholosiphy and Phenomenological Research*, vol. X, 1949–50, p. 172–197.

Husserl, E., *Erfahrung und Urteil*. Redigiert und hrsg. von L. Landgrebe. Hamburg, Claassen u. Goverts, 1948.

Jeans, Sir, J., Letter in *Nature*, vol. 148, 1941, p. 140–141, 255–257.

Johnson, M., *Science and the Meanings of Truth*. London, Faber & Faber, 1946.

Kant, I., *Critique of Pure Reason*, trans. by N. K. Smith. London, Macmillan, 1950.

Köhler, W., *Gestalt Psychology*. London, G. Bell, 1930.
- *The Place of Value in a World of Fact*. New York, Liveright, 1938.

Krasner, I., "Une Généralisation de la Notion de Corps," in *Journal de Mathématique*, 1er t., 1938.

Langer, Suzanne K., *Philosophy in a New Key*. Cambridge, Mass., Harvard University Press, 1942.

Lewis, C. I., *An Analysis of Knowledge and Valuation*. La Salle, Ill., Opencourt pub. co., 1946. (Carus Lectures).

Lovejoy, A. O., *The Revolt Against Dualism*. La Salle, Ill., Open Court pub. co., 1945. (Paul Carus Lectures).

McCrea, W. H., A Review of Eddington's *Fundamental Theory*, in *Mathematical Gazette*, vol. 31, 1947, p. 288–291.

Mariani, Jean., *Les Limites des Notions d'Objet et d'Objectivité*. Paris, Hermann, 1937. (Actualités Scientifiques et Industrielles, 519).

Meyerson, E., *De L'Explication dans les Sciences*. Paris, Payot, 1927.

Merleau-Ponty, M., *Phénoménologie de la Perception*. Paris, Gallimard, 1945

Milne, E. A., "The Inverse Square Law of Gravitation," in *Proceedings of the Royal Society*, A154, 1936, p. 22–52; A156, 1936, p. 62–85; A158, 1937, p. 324–348.

− "On the Origins of the Laws of Nature," in *Nature*, vol. 139, 1937, p. 997–999.

− A Review of Bridgman's *The Nature of Physical Theory*, in *Mathematical Gazette*, vol. 20, 1924, p. 340–342.

Plato, *Plato's Cosmology*, The Timaeus, trans. by F. M. Cornford. London, Routledge, Kegan Paul, 1937.

− *Plato's Theory of Knowledge*, The Theaetetus and the Sophist, translated by F. M. Cornford. London, Routledge & Kegan Paul, 1935.

Piaget, Jean., *La Construction du Réel chez l'Enfant*, 2 éd. Neuchatel, Delachaux, 1950.

Price, H. H., *Perception*, London, Methuen, 1932.

Russell, E., *The Analysis of Matter*. London, Allen & Unwin, 1935.

− *The Analysis of Mind*. London, Allen & Unwin, 1937.

− *Human Knowledge, Its Scope and Limits*. New York, Simon & Schuster, 1948.

− *An Inquiry into Meaning and Truth*. London, Allen & Urwin, 1940.

− *Problems of Philosophy*. London, Oxford University Press, 1946. (Home University Library).

Slater, Noel B., *The Development and Meaning of Eddington's 'Fundamental Theory.'* Cambridge, Eng., The University Press, 1957.

Stebbing, L., *Philosophy and the Physicists*. London, Methuen, 1937.

Walsh, W. H., *Reason and Experience*. Oxford, University Press, 1947.

Whitehead, A. N., *The Concept of Nature*. Cambridge, University Press, 1930. (Tarner Lectures for 1919).

− *Process and Reality, An Essay in Cosmology*. Cambridge, University Press, 1929.(Especially Part I, Chapter II,"The "Categoreal Scheme," pp. 24-42.

Whitrow G. J., *The Structure of the Universe, An Introduction to Cosmology*. London, Hutchinson, 1949.

− "The Epistemological Foundation of Natural Philosophy," in *Philosophy*, vol. XXI, 1946, p. 5–28.

Whittaker,E.T., *From Euclid to Eddington, A Study of the Conceptions of the External World*. Cambridge, University Press, 1949. (The Tarner Lectures).

− "Eddington's Theory of the Constants of Nature," in *Mathematical Gazette*, vol. 29 1945, p. 137–144.

# INDEX

28, 30, 32, 33, 34, 35, 39, 43, 44, 45,
46, 48, 49, 50, 52, 53, 54, 55, 60, 70,
76, 108, 114, 119, 120, 121, 122, 123,
124, 129, 141
world, perceptual, 5, 25
world, phenomenal, 23, 24, 25, 29, 30,
35, 92, 124

world, physical, 14, 15, 24, 25, 28, 29,
30, 31, 32, 34, 35, 37, 50, 56, 57, 60,
61, 77, 120, 121, 140
Whitehead, A. N., 36, 37, 38, 85
Whitrow, G. J., 67 n, 71, 72, 90, 98
Wittaker, E., 3, 85–86, 90, 91, 92, 93
Wittgenstein, L., 56

MARTINUS NIJHOFF — PUBLISHER — THE HAGUE

**Just out:**

# THE PHILOSOPHY OF SCIENCE

## OF

# *A. S. Eddington*

BY

### JOHN W. YOLTON

PRÉFACE PAR

F. GONSETH

*Zürich*

Eddington was a rare example of a scientist who was highly esteemed by his contemporaries for his scientific discoveries and ability and one whose reflections upon these discoveries were condemned thoroughly by both scientists and philosophers. He commanded sufficient prestige to be elected Gifford lecturer in the year 1927, and in 1937 he was further invited to give the Tarner lectures at Cambridge. After the Gifford lectures, he was definitely launched upon the path of speculation which brought him into conflict with many of the men in his own field. But it is interesting to note that he rounded of his series of publications by a return to a rigid formalism in the posthumous *Fundamental Theory*, a formalism which sought to sanctify the most radical of his philosophical interpretations of modern science by mathematical christening. Unlike many other scientists who have philosophized about science, Eddington cannot be properly evaluated until we have placed him in his historical context. His philosophy of science was an exercise in epistemology, and his epistemology was essentially and uniquely British. Many of the doctrines formulated and accepted by Russell reappear in Eddington, but these doctrines themselves have a long historical past ranging from Russell, Price, Broad and Moore to Locke, Berkeley, and Hume. The methodology and the ontology of this tradition have been preserved with few changes from Locke's day to Eddington's.

To those who have an interest in the British tradition of epistemology as well as to those who are curious about the many controversial claims of his philosophy of science, the present close analysis of Eddington's system in this historical context is worth-while.

*This study is a revised version of an essay submitted to the "Institut International des Sciences Théoriques" at Brussels for the Eddington contest it sponsored in 1951. This essay was awarded first prize by the "Institut" in January 1956.*

**About the author:**

B. A. and M. A. University of Cincinnati; Ph. D. Balliol College, Oxford University; taught at Johns Hopkins University and Princeton University; currently Associate Professor of Philosophy at Kenyon College, Gambier, Ohio; author of *John Locke and the Way of Ideas*, Oxford University Press; completing a two volume edition of Locke's *Essay Concerning Human Understanding* for Everyman's Library.

---

## CONTENTS

1960. XVI and 151 pages. roy. 8vo.

Guilders 11.50 = **$ 3.05** / **$ 3.20** postpaid

MARTINUS NIJHOFF — PUBLISHER — THE HAGUE

# LOGIC

### and the

# NATURE OF REALITY

*by*

### LOUIS O. KATTSOFF
University of North Carolina

"The book is valuable both because of what it attempts and how it proceeds. I recommend it to all who think that metaphysics is dead and as well to those who think that linguistic analysis has nothing to offer to the metaphysician. Both views could hardly be more mistaken. This book shows where and how they are mistaken by the best possible methods — by example. It produces a metaphysics which is very much alive, by using in crucial ways the tools of linguistic analysis." *The Personalist.*

1956. 247 pp.        Guilders 13.30 = **$ 3.55 / $ 3.75** postpaid

*By the same author:*

# PHYSICAL SCIENCE
# AND
# PHYSICAL REALITY

This is an extensive and well-designed work covering most of the topics that are customarily discussed in treatises on the philosophy of the physical sciences. The approach is mainly analytic, i.e. directed toward clarification of scientific concepts, in harmony with contemporary methods and studies. Nevertheless, the author tries throughout to connect his argument with broader philosophical questions. The treatment of the subject is such that the book is quite useful as a textbook for courses in the Philosophy of Science.

1957. VIII and 311 pp.
                        Cloth Guilders 17.75 = **$ 4.70 / $ 4.95** postpaid

**Edward G. Ballard,** Art and analysis. An essay toward a theory in aesthetics. 1957.
XV and 219 pp.　　　　　　　　Guilders 15.25 = $ 4.05 / $ 4.25 postpaid

**Harry M. Bracken,** The early reception of Berkeley's immaterialism 1710-1733.
1959. XII and 123 pp.　　　　　　Guilders 9.50 = $ 2.55 / $ 2.65 postpaid

**Gerd Brand,** Welt, Ich und Zeit. Nach unveröffentlichten Manuskripten Edmund
Husserls. 1955. XVI and 147 pp.　　Guilders 9.50 = $ 2.55 / $ 2.75 postpaid

**Constantin Brunner,** Der entlarvte Mensch. Herausgegeben und eingeleitet von
Lothar Bickel. 1951. XII and 205 pp.
　　　　　　　　　Cloth Guilders 12.— = $ 3.20 / $ 3.40 postpaid

**George S. Claghorn,** Aristotle's criticism of Plato's "Timaeus". 1954. XI and 149 pp.
　　　　　　　　　　　　Guilders 9.50 = $ 2.55 / $ 2.65 postpaid

**Joseph Cropsey,** Polity and economy. An interpretation of the principles of Adam
Smith. 1957. XII and 101 pp.　　Cloth Guilders 9.50 = $ 2.55 / $ 2.70 postpaid
= International Scholars Forum. A series of books by American scholars. Vol. 8.

**Erasmi opuscula.** A supplement to the Opera Omnia. Edited with introductions and
notes by Wallace K. Ferguson. 1933. XV and 373 pp.
　　　　　　　　　Cloth Guilders 15.— = $ 4.— / $ 4.25 postpaid

**James K. Feibleman,** Inside the great mirror. A critical examination of the philosophy
of Russell, Wittgenstein, and their followers. 1958. 228 pp.
　　　　　　　　　Cloth Guilders 19.— = $ 5.05 / $ 5.30 postpaid

**Eugen Fink,** Zur ontologischen Frühgeschichte von Raum, Zeit, Bewegung. 1957. XI
and 247 pp.　　　　　　Cloth Guilders 15.75 = 4.20 / $ 4.40 postpaid

**Eugen Fink,** Alles und Nichts. Ein Umweg zur Philosophie. 1959. VIII and 250 pp.
　　　　　　　　　Cloth Guilders 15.75 = $ 4.20 / $ 4.40 postpaid

**For Roman Ingarden.** Nine essays in phenomenology. 1959. VIII and 179 pp.
　　　　　　　　　　Guilders 15.25 = $ 4.05 / $ 4.25 postpaid

**Carl H. Hamburg,** Symbol and reality: Studies in the philosophy of Ernst Cassirer.
1956. IX and 172 pp.　　　　Cloth Guilders 12.40 = $ 3.30 / $ 3.50 postpaid

**R. J. Henle S.J.,** Saint Thomas and Platonism. A study of the Plato and Platonici texts
in the writings of Saint Thomas. 1956. XXIII and 487 pp.
　　　　　　　　　Cloth Guilders 30.— = $ 8. — / $ 8.25 postpaid

**G. Heymans,** Gesammelte kleinere Schriften zur Philosophie und Psychologie.
1927. 3 vols. I: Erkenntnistheorie und Metaphysik. XI and 477 pp. With portrait
and 2 plates. II: Allgemeine Psychologie, Ethik und Aesthetik. VI and 615 pp.
With textfigures. III: Spezielle Psychologie. VIII and 630 pp. With textfigures.
　　　　　　　　　　Guilders 45.— = $ 11.95 / $ 12.75 postpaid

**Kurt Hildebrandt,** Leibniz und das Reich der Gnade. 1953. VIII and 505 pp.
　　　　　　　　　Cloth Guilders 26.50 = $ 7.05 / $ 7.35 postpaid

**Edmund Husserl,** Cartesian Meditations. An introduction to phenomenology.
Translated by Dorion Cairns. 1960. XII and 157 pp.
Paper bound, 5¹/₂ by 8¹/₂. (College textbook) Guilders 9.50 = $ 2.50 / $ 2.70 postpaid
Cloth bound, 6¹/₄ by 9¹/₂.　　　　　　Guilders 13.25 = $ 3.50 / $ 3.70 postpaid

**Husserliana - Edmund Husserl, Gesammelte Werke.**

　I. Cartesianische Meditationen und Pariser Vorträge. Herausgegeben und einge-
　　leitet von S. Strasser. 1950. XXXI and 244 pp.
　　　　　　　　　　　　Guilders 10.— = $ 2.65 / $ 2.90 postpaid
　　　　　　　　　Cloth Guilders 12.50 = $ 3.35 / $ 3.55 postpaid

　II. Die Idee der Phänomenologie. Fünf Vorlesungen. Herausgegeben und einge-
　　leitet von Walter Biemel. 2. Auflage. 1958. XII and 94 pp.
　　　　　　　　　　　　Guilders 4.50 = $ 1.20 / $ 1.40 postpaid
　　　　　　　　　Cloth Guilders 6.60 = $ 1.75 / $ 1.95 postpaid

　III. Ideen zu einer reinen Phänomenologie und phänomenologischen Philosophie.

**Husserliana - Edmund Husserl, Gesammelte Werke** (continued)

Erstes Buch: Allgemeine Einführung in die reine Phänomenologie. Herausgegeben von Walter Biemel. XVI and 483 pp.

Guilders 20.— = $ 5.30 / $ 5.60 postpaid
Cloth Guilders 23.50 = $ 6.25 / $ 6.55 postpaid

IV. Id. Zweites Buch: Phänomenologische Untersuchungen zur Konstitution. Herausgegeben von Marly Biemel. 1952. XX and 426 pp.

Guilders 19.— = $5.05 / $ 5.35 postpaid
Cloth Guilders 22.50 = $ 6.— / $ 6.30 postpaid

V. Id. Drittes Buch: Die Phänomenologie und die Fundamente der Wissenschaften. Herausgegeben von Marly Biemel. 1952. VI and 164 pp.

Guilders 6.75 = $ 1.80 / $ 2.— postpaid
Cloth Guilders 10.— = $ 2.65 / $ 2.90 postpaid

VI. Die Krisis der europäischen Wissenschaften und die transzendentale Phänomenologie. Eine Einleitung in die phänomenologische Philosophie. Herausgegeben von W. Biemel. 1954. XXII and 559 pp.

Guilders 23.50 = $ 6.25 / $ 6.55 postpaid
Cloth Guilders 27.50 = $ 7.30 / $ 7.60 postpaid

VII. Erste Philosophie (1923/24). 1. Teil: Kritische Ideengeschichte. Herausgegeben von Rudolf Boehm. 1956. XXXIV and 468 pp.

Guilders 21.50 = $ 5.70 / $ 6.— postpaid
Cloth Guilders 26.25 = $ 7.— / $ 7.25 postpaid

VIII. Erste Philosophie (1923/24). 2. Teil: Theorie der phänomenologischen Reduktion. Herausgegeben von Rudolf Boehm. 1959. XLIII and 594 pp.

Guilders 26.75 = $ 6.40 / $ 6.70 postpaid
Cloth Guilders 32.— = $ 8.50 / $ 8.85 postpaid

*In preparation:*

IX. Phänomenologische Psychologie.

**Georg G. Iggers,** The cult of authority. The political philosophy of the Saint-Simonians. A chapter in the intellectual history of totalitarianism. 1958. VIII and 210 pp.

Cloth Guilders 14.25 = $ 3.80 / $ 4.— postpaid

**Oliver A. Johnson,** Rightness and goodness. A study in contemporary ethical theory. 1959. VII and 163 pp. Cloth Guilders 15.25 = $ 4.05 / $ 4.25 postpaid

= International Scholars Forum. A series of books by American scholars. Vol.. 13

**Kwee, Swan Liat,** Methods of comparative philosophy. 1953. X and 217 pp.

Guilders 7.90 = $ 2.10 / $ 2.30 postpaid

**Lachelier.** — The philosophy of Jules Lachelier. "Du fondement de l'induction", "Psychologie et métaphysique", "Notes sur le Pari de Pascal" together with Contributions to "Vocabulaire technique et critique de la philosophie" and a selection from his letters, translated and introduced by EDWARD G. BALLARD. 1960. XV and 118 pp.  Guilders 8.50 = $ 2.25 / $ 2.45 postpaid

**B. Landheer,** Mind and society. Epistemological essays on sociology. 1952. XII and 112 pp.  Guilders 5.50 = $ 1.45 / $ 1.65 postpaid

**B. Landheer,** Pause for transition. An analysis of the relation of man, mind and society. 1957. V and 284 pp.  Guilders 21.— = $ 5.60 / $ 5.80 postpaid

**J. H. Leopold,** Ad Spinozae opera posthuma. 1902. VIII and 92 pp.

Guilders 2.— = $ 0.55 / $ 0.60 postpaid

**Philip Merlan,** From Platonism to Neoplatonism. Second revised and enlarged edition. Guilders 17.— = $ 4.55 / $ 4.80 postpaid. *(In preparation)*

# MARTINUS NIJHOFF — PUBLISHER — THE HAGUE

**Phaenomenologica** - Collection publiée sous le patronage des Centres d'Archives-Husserl.

    **Comité de Rédaction:** Président: H. L. Van Breda (Louvain). Membres: M. Farber (Buffalo), E. Fink (Fribourg en Brisgau), J. Hyppolite (Paris), L. Landgrebe (Cologne), M. Merleau-Ponty (Paris), P. Ricoeur (Paris), K. H. Volkmann-Schluck (Cologne), J. Wahl (Paris). Secrétaire: J. Taminiaux (Louvain).

1. **Eugen Fink,** Sein, Wahrheit, Welt. Vor-Fragen zum Problem des Phänomen-Begriffs. 1958. VIII and 156 pp. Cloth Guilders 12.50 = **$ 3.35 / $ 3.50** postpaid

2. **Husserl et la pensée moderne / Husserl und das Denken der Neuzeit.** Actes du deuxième Colloque International de Phénoménologie, Krefeld, 1-3 novembre 1956, édités par les soins de H. L. Van Breda et J. Taminiaux / Akten des zweiten Internationalen Phänomenologischen Kolloquiums, Krefeld, 1.-3. November 1956, herausgegeben von H. L. Van Breda und J. Taminiaux. 1959. X and 250 pp.     Cloth Guilders 16.— = **$ 4.25 / $ 4.50** postpaid

3. **J.-Cl. Piguet,** De l'esthétique à la métaphysique. 1959. VI and 294 pp.     Cloth Guilders 24.— = **$ 6.40 / $ 6.60** postpaid

4. **Edmund Husserl, 1859-1959.** Recueil commémoratif publié à l'occasion du centenaire de la naissance du philosophe. 1959. XI and 306 pp. With portrait.     Cloth Guilders 20.— = **$ 5.30 / $ 5.60** postpaid

    *In preparation:*

    **A. Roth,** Edmund Husserls ethische Untersuchungen.

    **Herbert Spiegelberg,** The phenomenological movement. A historical introduction. 2 volumes.

    **M. D. Cairns,** Guide for reading Husserl.

**M. W. Rombout,** La conception stoïcienne du bonheur chez Montesquieu et chez quelques de ses contemporains. 1958. 125 pp.
    Guilders 12.50 = **$ 3.35 / $ 3.50** postpaid
= Leidse Romanistische Reeks, IV.

**Lazare Saminsky,** Physics and metaphysics of music and essays on the philosophy of mathematics. (A green philosopher's peripeteia. Physics and metaphysics of music. The roots of arithmetic. Critique of new geometrical abstractions. The philosophical value of science.). 1957. 151 pp. Cloth Guilders 10.45 = **$ 2.80 / $2.95** postpaid

**B. de Spinoza,** Opera quotquot reperta sunt. Recognoverunt J. van Vloten et J. P. N. Land. Editio tertia. 1914. 4 vols. XXXII and 1104 pp.
    Cloth Guilders 20.— = **$ 5.30 / $ 5.70** postpaid

**Spinoza Mercator et Autodidactus** - Oorkonden en andere authentieke documenten betreffende des wijsgeers jeugd en diens betrekkingen, verzameld door A. M. Vaz Diaz. Uitgegeven en toegelicht in overleg met W. G. van der Tak. 1932. XII and 69 pp. With 13 facs.     Guilders 10.— = **$ 2.65 / $ 2.90** postpaid

**Joan Stambaugh,** Untersuchungen zum Problem der Zeit bei Nietzsche. 1959. XX and 235 pp.     Cloth Guilders 21.50 = **$ 5.70 / $ 5.95** postpaid

**Tulane Studies in Philosophy, Volumes I–VIII,** 1952 - 1959.
    Per volume Guilders 7.60
Purchasers in the U.S.A. should order the volumes of Tulane Studies in Philosophy for $ 2.— per volume, plus postage, from the Department of Philosophy, Tulane University, New Orleans 18, Louisiana.

*The prices quoted in US $ are based on the*
*rate of exchange: 1 guilder =: $ 0.266 (September 1960)*
*They are subject to change without notice*